W9-CPJ-042

Pilot's WEATHER

– a commonsense approach to meteorology

Pilot's
WEATHER

– a commonsense approach to meteorology

Brian Cosgrove

Plymouth Press

Dedication

To John Bannister, Julian Doswell and Peter Coles –
a truly patient trio

Copyright © 1999 Brian Cosgrove

First published in the United States by Plymouth Press
Previously published in the UK in 1999
by Airlife Publishing Ltd

British Library Cataloguing-in-Publication Data
A catalogue record for this book
is available from the British Library

ISBN 1-882663-41-1

All rights reserved. No part of this book may be
reproduced or transmitted in any form or by any means,
electronic or mechanical including photocopying,
recording or by any information storage and retrieval
system, without permission from the Publisher in
writing.

The information in this book is true and complete to
the best of our knowledge. All recommendations are
made without any guarantee on the part of the
Publisher, who also disclaims any liability incurred in
connection with the use of this data or specific details.

Printed in Hong Kong

Photo Credits:
Tom Bradbury pages 37, 39, 70, 78 top/bottom,
88 bottom, 103, 140
Julian Doswell page 102
Terry Skuce page 67
Jim Ellis page 106

Distributed in the US and Canada by Plymouth Press
101 Panton Road, Vergennes, VT 05491.

To order additional copies or for a free catalog, call
1-800-350-1007

CONTENTS

INTRODUCTION

Pilot's Weather aims to provide the meteorological 'know-how' needed by you as a student pilot – and just maybe some existing pilots! It will be relevant whether your aircraft is a light aeroplane, sailplane, hang/para glider, ultralight, microlight, powered parachute or whatever.

Simplicity is the intention as far as possible; achieved through adopting a logical progression of the events to be absorbed and using numerous pictures of the clouds and weather patterns you will come across.

Past experience has shown that even at basic level there are a few whose curiosity likes to go that little bit deeper at the outset. Where such occasions arise, the text will be found in italics placed in a pink tinted box and can be ignored if your only wish is to learn the basics.

Should the day come when you want to extend your pilot status up to commercial or airline standard, or you wish to become a highly skilled sailplane pilot achieving a Gold C with diamonds, there are plenty of other books available to take you that stage further. In the meantime *Pilot's Weather* will ensure you at least have a basic rock upon which to build the advanced knowledge that any future aspirations may demand.

Meteorological terminology and units of measurement are steadily becoming uniform internationally, but currently in the USA there are differences. For students resident or visiting the United States to learn to fly, every effort has been made to identify the major differences in terminology and procedures. Also, since the 1st January 1996 changes have been taking place – particularly in the reporting of weather – to become more in line with the international approach. To cover this transient period, terms will be duplicated in this book where necessary.

Apart from the basic principles of meteorology, the book highlights the weather problems with which you can be faced during your flying life – it is *not* simply confined to your passing an examination.

You may become very competent in other aspects of flight, *BUT* – can any of them be more crucial than a sound familiarity with the environment through which you are prepared to fly yourself and your passengers? When putting your knowledge of met into practice always err on the safe side and use the information available to you in the form of current reports and forecasts by phone, fax, radio or computer.

It is better to be on the ground wishing
you were in the air than in the air
wishing you were on the ground!

Brian Cosgrove

CHAPTER ONE

THE ATMOSPHERE

Quite simply the atmosphere is the air surrounding our planet Earth – mercifully kept in place by the force of gravity.

It is primarily composed of 78% nitrogen, 1% of other gases and, fortunately, 21% oxygen for which we should be truly grateful. Apart from life as we know it being unable to survive without oxygen, combustion would not be possible for engines to function.

As the force of gravity is greater nearest to its source, so the atmosphere is more concentrated near our planet's surface.

There are four basic factors that affect the weather and flight.

They are:

DENSITY	– the weight of a given volume of air.
PRESSURE	– the weight of a given column of air in the atmosphere.
TEMPERATURE	– the warmth of the atmosphere.
HUMIDITY	– the moisture content of the atmosphere.

For occasions when uniformity or a working standard is required, the **International Civil Aviation Organisation (ICAO)** has agreed on an **International Standard Atmosphere (ISA).** This lays down standards for density, pressure and temperature based on a defined sea level — more on this later.

You must remember that these standards are but a 'yardstick' and they are purely theoretical. Rarely would the conditions in the atmosphere ever coincide with the standards all at the same time.

How these ISA standards play a part in aviation will be made clear as we progress.

CHAPTER TWO

DENSITY

Air has weight, and **density** is simply the weight of the number of molecules of air present in a given **'parcel'** or volume of air at a given time. (The word parcel is frequently used in meteorology as no one has yet thought of a better alternative!)

Density is greatest at the surface; it decreases with height as the air thins out until in outer space it is no longer relevant.

Apart from a decrease with height, density is also affected by warmth. When a parcel of air of a given volume is heated it becomes thinner as the molecules spread out and their number becomes less than in the original parcel.

Conversely, when a volume of air is cooled it becomes denser.

Density also decreases as the moisture content in the air increases, but the effect is relatively minimal.

The ISA standard density at sea level (excepting the Dead Sea!) is 1.225 kilograms per cubic metre (or 1225 grams) or 0.764 lb per cubic foot, where density is defined as 100%. In other words, this would be the defined average weight of all the air molecules in a volume of one cubic metre of air at sea level.

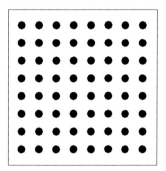

Unlike the other basic factors we shall cover, it cannot be readily measured by instrumentation – its value at a given time requires calculation.

Density plays an important part in an aircraft's performance; in fact an aircraft's design limitations are based on the ISA standards. However, as already mentioned, reality rarely matches laid down standards; they are but yardsticks.

The effect of density below the established 100% at 1.225 kg per cubic metre has a definite bearing on an aircraft's stated 'book' performance. This area we shall discuss after you have digested the aspects of pressure, temperature and humidity.

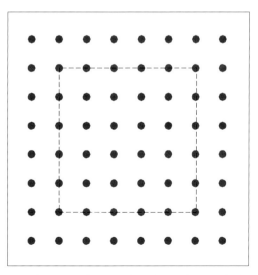

Warmed air expands – notice how the number of molecules in the original volume has decreased

CHAPTER THREE

PRESSURE

WHAT IT IS

Pressure is the weight of a given *column* of air in the atmosphere measured at the earth's surface. It decreases with height until becoming nil when air ceases to exist. You may think that the reference to pressure being weight means it is the same as density — this is not so.

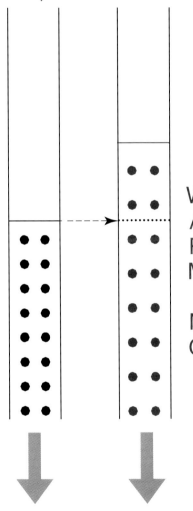

The surface pressure need not decrease with a decrease in density

Imagine warming a given parcel of air in a tube. This air would expand and spread itself farther up the tube so that the *column* in the tube would become greater in length. But take the volume of the *original parcel* of air in the longer column and you will find it is lighter (less dense) than it was before the warming took place. The same amount of air remains in the column overall so the pressure at the bottom stays the same.

WARMING

MEASURING PRESSURE

The unit by which atmospheric pressure is measured in the UK has for many years been the **millibar (mb)**. The ICAO has now agreed to a change of unit to that of the **hectoPascal (hPa)**. The change is rather academic in that 1 mb equals 1 hPa so it is a case of "a rose by any other name"!

However, the millibar will continue to be used in the UK for operational purposes for the foreseeable future; the hectoPascal will be confined only to those occasions when reference is being made to ICAO International Standard Atmosphere as a definition in itself.

In the light of this present situation we will naturally refer to millibars (mbs) in this book with possibly the occasional insertion of hPa as a reminder for the future.

In the United States of America, measurement of pressure is currently in **inches (in.Hg)** – based on the length of a column of mercury in circumstances about to be described.

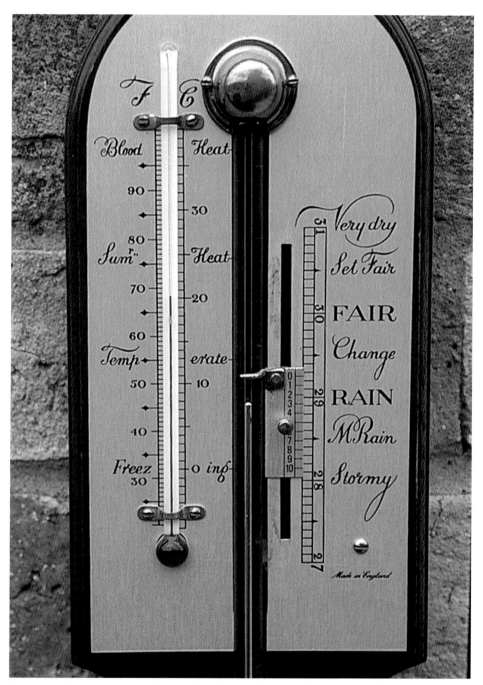

The mercury barometer consists of a long thin tube extending upwards from a reservoir of mercury. A tiny hole in the reservoir allows air pressure to enter and exert its influence on the mercury. The above is an example from where the inch is the unit of measurement

Readily recognisable with its clock face and scale marked with a range of choices from 'Set Fair' to 'Stormy', or simply a plain dial showing measurement only. A perfectly reasonable instrument can be purchased for a very modest outlay. There is usually an adjustable hand which can be set over a reading at a given time so that future changes are readily observed

The instrument used to measure pressure is the **barometer**, of which there are two main types.

Mercury barometer

This consists of a transparent (usually glass) tube in which a column of mercury extends up from a reservoir at the base of the tube. A tiny inlet hole allows air pressure to enter the reservoir, and the column rises or falls according to the increase or decrease in outside air pressure. The tube is graduated with a scale that is sometimes fitted with a vernier to permit the maximum accuracy in reading off either millibars or inches.

Aneroid barometer

This type consists of a sealed concertina-like drum wherein a fixed pressure exists. As outside pressure increases or decreases the drum contracts or expands, and its movement is transmitted to a dial or a digital read-out. Tradition in the UK still sees the inch measurement on many household barometers, sometimes accompanied by millibars on later models.

The reading can also be transmitted to a revolving drum where a continuous picture of pressure changes can be seen. This type of aneroid barometer is known as a **barograph.**

Here you can see the sealed, concertina-like drum and the linkage associated with an aneroid barometer. In this case the linkage transmits a continuous reading to a chart

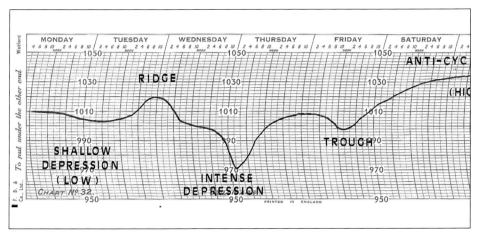

The barograph can be very valuable for amateur weather forecasting

Conversion from millibars to inches and vice-versa is as follows:

$$1 \text{ millibar} = 0.02953 \text{ in.Hg}$$
$$\text{and } 1 \text{ in.Hg} = 33.8653 \text{ mb}$$

For example   1000 mb   ÷   33.8653   =   29.53 in.Hg
            or 29.53 in.Hg   ÷   0.02953   =   1000 mb.

Those who enjoy calculations should find that 27 ft per 0.02953 in.Hg equates to 30 ft being the equivalent of 0.032811 in.Hg – hence the 0.3 in.Hg per 30 ft referred to above makes it all the more simple.

At around 20,000 ft in the more rarefied air the change becomes approximately 50 ft per unit of 1 millibar or 0.02953 in.Hg, and increases to approximately 75 ft per unit around 30,000 ft due to the markedly reduced density at that height.

PRESSURE CHANGES WITH HEIGHT

The ISA standard for a change in height is 27 ft per hectoPascal (mb) – the equivalent of 0.02953 in.Hg. However, the generally accepted change is the round figure of **30 ft per millibar** which equates to **0.03 in.Hg** and this combination we will use when required.

UNIFORMITY IN MEASUREMENT

Taking the accepted decrease in pressure with height at 30 feet per millibar (0.03 in.Hg), given a surface pressure of 1000 mb (29.53 in.Hg), a reading taken at the top of a 300 ft (100 metres) building would be 990 mb (29.23 in.Hg) for the same location.

You will realise that if pressure readings were taken at stations located at varying heights it would be impossible to obtain a meaningful pressure pattern at any given time.

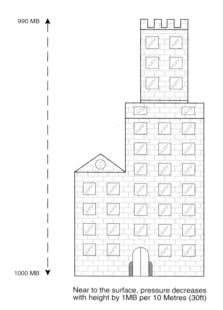

Near to the surface, pressure decreases with height by 1MB per 10 Metres (30ft)

990 MB

1000 MB

100 METRES

Uniformity is obtained by all pressure readings being converted to sea level, with even tidal movement being taken into account, thus ending up with the term **Mean Sea Level (MSL)**.

The ISA MSL standard is **1013.25 mb** (hPa) or **29.92 in.Hg**.

Pressure readings, together with other weather data, are noted every day at regular intervals throughout the world and transmitted to main centres. Here the pressures are reproduced on charts with lines joining places of equal MSL pressure. These lines are known as **isobars** and appear similar to contours on a map.

The difference between top and bottom would be 10 mb

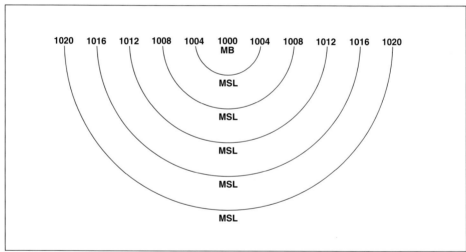

1020 1016 1012 1008 1004 1000 MB 1004 1008 1012 1016 1020

MSL

MSL

MSL

MSL

MSL

The main pressure systems

PRESSURE SYSTEMS

Isobars produce a pattern which depicts various pressure systems.

HIGH	– A centre of high pressure known as an **anticyclone**.
RIDGE	– A wedge or tongue of high pressure extending out from the centre of an anticyclone.
LOW	– A centre of low pressure known as a **depression**.
TROUGH	– A wedge or tongue of low pressure extending out from the centre of a depression.
COL	– The neutral area between two highs and two lows.

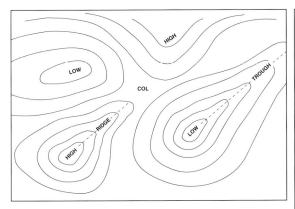

For operational reasons the pressure pattern in the upper air is depicted by contours. These lines join places where a given pressure is at the same height – this time the principle is the same as that of contours on a map, with the selected pressure level, say the 500 mb (13.65 in.Hg) level, shown in metres and/or feet.

Isobars joining places of equal MSL pressure

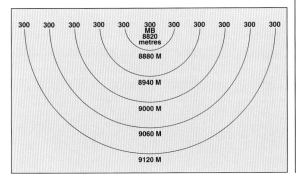

Contours tell the same story as isobars but the location of the upper air pressure systems they depict in terms of centres of high and low pressure can differ in location from those depicted by isobars at sea level.

Contours joining places of equal height above MSL for a given pressure level

WORLD PRESSURE DISTRIBUTION

There are appropriate general areas of high and low pressure in belts surrounding the Earth.

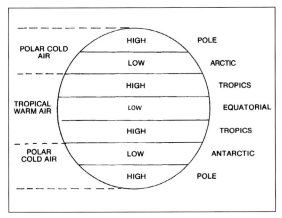

The world pressure distribution

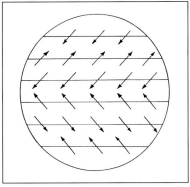

The rotation of the earth causes the flow to deflect and not move directly from high to low as you might expect

CHAPTER FOUR

THE ALTIMETER

Excluding the radio version, the altimeter is included in this book at this stage because of its direct relationship with meteorology in relying upon atmospheric pressure for its operation. It is the instrument which indicates the height of an aircraft above a pre-selected surface level and is nothing more than a sensitive aneroid barometer. But in this case it is calibrated to read feet, instead of millibars or inches, on the basis of the 1 mb (0.03 in.Hg) change in pressure equating albeit to a change of 30 ft in terms of height.

Atmospheric pressure effect on the aneroid, known as **static pressure**, is fed to the instrument from an aperture set into the fuselage side at right angles to any airflow.

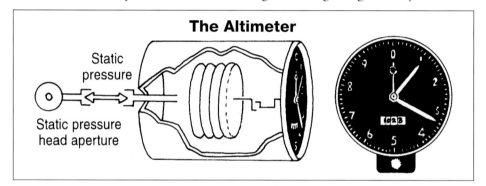

The Altimeter

Static pressure

Static pressure head aperture

Air pressure enters static head and on to the aneroid drum, causing it to compress and expand accordingly

The altimeter can have a three-handed dial. The first and largest hand will indicate feet in hundreds; the second, smaller hand will show thousands and the third, with possibly a ring or an inverted triangle on it, depicts tens of thousands of feet. This three-hand format can lead to inadvertent mis-reading and has done so on a number of occasions in the past, so be particularly careful in your interpretation. Some altimeters also give a digital read-out.

Below the face of the instrument is a knob which enables a selected pressure setting to be entered at any given time on the ground or in flight. Calibration of the setting scale can be in millibars (mb) or inches (in.Hg) or both. Unlike the barometer, stationary at a reporting station, the altimeter is very much a mobile entity which is subject to many variables.

Take one such situation – atmospheric pressure is constantly changing, so the altimeter has to be adjusted in order to read correctly at the time.

Eg: The altimeter is reading 0 ft at, say, a surface pressure of 1000 mb (29.53 in.Hg). The pressure drops to 990 mb. (29.23 in.Hg) which, at the rate of 1 mb (0.03 in.Hg) per 30 ft, is the equivalent of 300 ft.
The altimeter will now read 300 ft although the aircraft is still on the ground.
Re-setting to 990 mb (29.23 in.Hg) will see the altimeter once again reading 0 ft.

A basic altimeter with a millibar setting scale

This altimeter has two scales – the left one is in millibars – the one on the right is in inches

On a long cross-country flight, *en route* pressure changes can take place and the pressure can be totally different on arrival at the destination. If on radio, this can be checked and adjustments made to the altimeter setting; without radio the only alternative is to seek information on possible changes prior to take-off.

Before going into detail on how defined laid-down pressure settings help to control this situation, first take a look at how being unaware of such a change of pressure can affect a planned flight path to a destination.

For simplicity's sake, the following example, albeit exaggerated in terms of pressure change, presupposes both take-off point and destination to be at sea level.

Eg: • Airfield pressure setting at take-off shows 1020 mb. (30.12 in.Hg).
   • Altimeter reads 0 ft.
   • Aircraft climbs out and sets course at 2000 ft.
   • During the flight the pilot notices height 'increasing' and descends to maintain 2000 ft on the altimeter.
   • At destination on landing, being the same height as at take-off, the altimeter reads *990 ft.*
   • Pressure has decreased by 33 mb (0.97 in.Hg) to 987 mb (29.15 in.Hg) – equivalent to a false increase in height of 990 feet.
   • Actual height on arrival overhead had been *1010 ft,* not 2000 ft.
   • There had been a *sloping* downward flight path due to the decrease in pressure from take-off point to destination.

It needs little imagination to realise that, unless there is a safeguard in place, any flight path that slopes downward can pose serious problems to a pilot flying in poor visibility or 'blind' on instruments – particularly if high ground is on the route and a safety height has not been built in to the planned flight path.

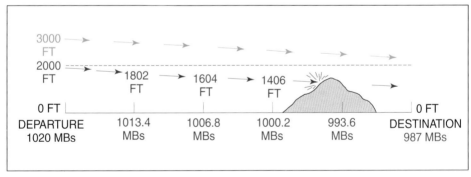

When flying towards a centre of low pressure the altimeter can give a false higher altitude reading

Again, the likelihood of the height above mean sea level (**AMSL**) of the take-off and destination points being identical is rare indeed. In the above example, if the destination had been 300 ft above the take-off point, the height on arrival overhead would have been *710 ft,* not 1010 ft, or the 2000 feet that the pilot expected.

To overcome such situations there is a range of defined altimeter settings of which there are four in the UK. In the USA there are only two settings; we will take the more detailed UK procedure first.

ALTIMETER SETTINGS – UNITED KINGDOM

In the UK each of the four settings forms a **datum** to which you must adhere in given situations. They ensure that aircraft in the airfield circuit or in *en route* flight are flying to a common datum.

QFE
- Pressure setting at airfield level.
- Set in proximity to home or destination airfield when due to land or during circuit flying. Also may need to be used in proximity to another airfield or MATZ.
- The reading from this setting is officially referred to as **height** and registers 0 ft on landing.
- QFE is currently becoming less used by some large transport aircraft.

QNH
- Pressure setting at MSL.
- Set just prior to take-off when going cross-country.
- Used mainly in flight across country.
- The reading from this setting is officially referred to as **altitude** and registers the elevation of the airfield above sea level on landing.

REGIONAL QNH
- Pressure setting at a given QNH for a specified region.
- It is based on the lowest forecast QNH for the region to ensure safer terrain clearance; it is amended at regular intervals.
- This reading is also known as **altitude**.
- Used *en route* when away from the airfield vicinity as an alternative to QNH.
- The various altimeter setting regions are depicted on aeronautical charts.
- It also assists with *en route* aircraft separation when positions are reported to a controller.

STANDARD PRESSURE SETTING (SPS)
- This pressure setting is the defined ISA Standard Pressure Setting (SPS) expressed in millibars as 1013.25 at mean sea level (MSL). In practice the nearest you can probably get on the altimeter scale will be **1013 mb** or the mark which appears on some scales for this setting.
- The SPS is set on passing a stated **transition altitude** after take-off (more later).
- The altitude reading at the SPS is known as the **pressure altitude** but it is expressed as a **Flight Level (FL)** where the hundreds (00s) are deleted from the pressure altitude reading. For example, 15,500 ft would be FL 155.
- Flight levels are established at 500 ft intervals and always relate to the SPS of 1013 mb. In this way adequate aircraft separation is ensured – essential when flight is under **IFR (Instrument Flight Rules)**.

What may give rise to concern is grasping the mechanics of the differences the two settings (SPS and QNH) can portray in the messages they give out. **On this aspect you are far from alone – such concern has been experienced by many before you!**
First a brief repetition:

- SPS is the defined *fixed* value of 1013 mb by which altimeters are set to ensure uniformity of operation when required. It is a datum in the form of the ISA-defined standard 'mean sea level' MSL pressure.
- QNH is the actual pressure at MSL at any given time and is a *variable* – it is constantly changing.

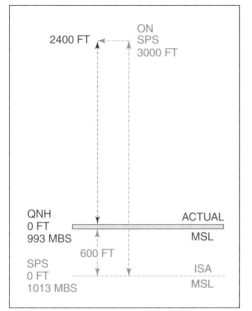

A QNH value below 1013 mb (29.92 in.Hg) will produce a QNH altitude reading below the SPS pressure altitude

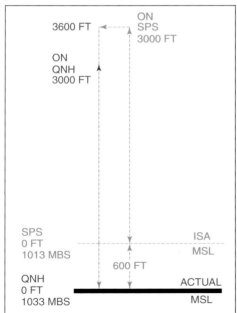

A QNH value above 1013 mb (29.9 in.Hg) will produce a QNH altitude reading above the SPS pressure altitude

Now we will look at the relationship between the two and its effect. Apart from looking at the diagrams provided, I would suggest you follow along with your own drawings to get the feel of it.

(a)　When the altimeter is set at the SPS of 1013 mb, if the QNH figure is **less** than the ISA MSL datum then the **real** altitude of the aircraft is **less** than the altimeter reading.

For example, you can see from the diagram, when the QNH figure is **lower** than the SPS of 1013 mb the actual surface is **above** the ISA MSL datum. In this case to the tune of 600 ft (1013 – 993 = 20 mb x 30 ft = 600 ft).

So, a reading of 3000 ft on an altimeter set to the SPS of 1013 mb will show an altitude of 2400 ft when the setting is changed to say a QNH of 993 mb.

(b)　Conversely, when the altimeter is set at the SPS of 1013 mb, if the QNH figure is **more** than the ISA MSL datum then the **real** altitude of the aircraft is **more** than the altimeter reading.

This time you can see from the diagram that when the QNH figure is **higher** than the SPS of 1013 mb the actual surface is **below** the ISA MSL datum. Again, in this case, to the tune of 600 ft (1013 – 993 = 20 mb x 30ft = 600 ft).

For example, a reading of 3000 ft on an altimeter set to the SPS of 1013 mb will show an altitude of 3600 ft when the setting is changed to say a QNH of 1033 mb.

A useful rule. When the millibar figure to be set at a change is **less** than the existing setting, the altimeter readings shown at the change will be **less**.
Or simply, **millibars set to less – indicate feet less.**
Conversely, **millibars set to more – indicated feet more**

We now go on to discuss the procedure when you have to change from QNH to the Standard Pressure Setting of 1013 mb.

Transition level/layer

When outside controlled airspace in the UK the **transition altitude**, referred to earlier, is 3000 ft. Inside controlled airspace it can vary and, when preparing your flight plan, you must ascertain the appropriate transition altitudes for any controlled airspace along your route and/or your destination. It is at this altitude that you will change from the QNH to the SPS setting of 1013 mb.

The **transition level** is the first 500 ft flight level (FL) that is reached *after* climbing through the transition altitude. The transition level is also the level on descent where a change back from the SPS to QNH is made which can be followed by a further change to QFE if about to enter an airfield circuit.

We now take a look at examples of changing from QNH to the SPS to attain the correct transition level in controlled airspace. In this case the transition altitude is 4000 ft, and bear in mind that flight levels are at 500 ft intervals. Again, in all cases, the first flight level after transition altitude is the defined transition level.

NOTE:

To help you grasp the next two examples visually, an altitude scale of which you can take photocopies, is to be found at the end of this chapter.

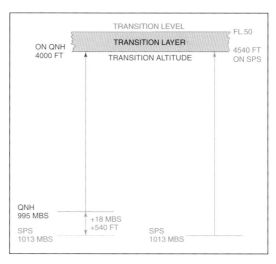

Example 1.

When the QNH at, say, 995 mb is *less* than the fixed SPS of 1013 mb, then on changing to 1013 mb **(more millibars)** at a transition altitude of, say, 4000 ft, the altimeter will now show a *higher* reading **(more altitude)** of 4540 ft. The first available flight level will be FL 50.

Example 2.

When the QNH at, say, 1023 mb is *greater* than the fixed SPS of 1013 mb, then on changing to 1013 mb **(less millibars)** at a transition altitude of, say, 4000 ft, the altimeter will show a *lower* **(less altitude)** reading of 3700 ft. The first available flight level will be FL 40.

The space between the QNH transition altitude and the transition level flight level is known as the **transition layer**.

Invariably the QNH will differ from the SPS, which in turn

ABOVE & BELOW Differences between the QNH and the SPS can vary the transition level for the same QNH altitude

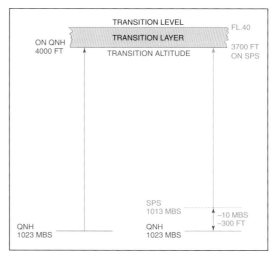

means that rarely will a QNH altitude match a flight level. This shows how care must be taken in how you interpret flight levels.

Consider the situation when you are approaching a Control Area (CTA) at 4200 ft, you are on a QNH of 990 mb and the CTA has its base at FL 45 (a pressure altitude of 4500 ft when related to the SPS at 1013 mb).

The difference of 23 mb **less**, between the QNH and the SPS is 690 ft (23 mb x 30ft), which in this case means that FL 45 is lower – actually only 3810 ft (4500 – 690) above the AMSL datum that time.

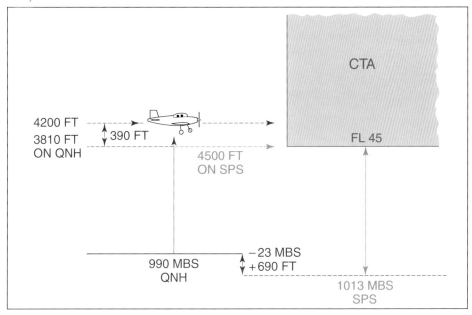

A QNH below the SPS could lead to an infringement of controlled airspace

The result could be an infringement of the CTA if you continued your flight with the altitude at 4200 ft on your QNH setting.

It is one thing to infringe controlled airspace but quite another to 'infringe' high gound due to being unaware of your real altitude as opposed to the indicated altitude.

For example:

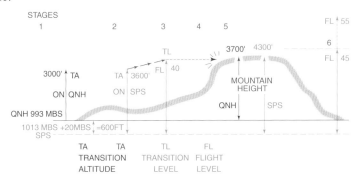

Stages
1. You reach a transition altitude of 3000 ft on a QNH of 993 mb.
2. On changing to the SPS of 1013 mb your altimeter is now reading 3600 ft. (**More millibars – more altitude** because 993 to 1013 = + 20 mb x 30 ft = + 600 ft).
3. The transition level being the first 500 ft Flight Level after transition altitude you now climb to FL 40 (4000 ft on SPS).
   You are aware of high ground ahead reaching up to 3700 ft above MSL so with good visibility and 4000 ft on the altimeter you may well press on quite happily.
4. **BUT**, at FL 40 the real altitude is actually only 3400 ft.
5. Without careful pre-flight planning and running into poor visibility *en route* you may easily plough into the mountain – a type of cloud know to cynical pilots as *Cumulus Granitus.*
6. Even selecting FL 45 in your pre-flight planning would be far too low a figure. FL 55 or higher would be wiser.

At the risk of repetition, be warned when the QNH is **less** than the ISA MSL datum of 1013.25 mb, the FL will be **lower** than its stated figure on your map/chart.

You may have noticed that the emphasis has been on situations where the QNH is lower than the SPS. When it is **greater** there is no problem; you will always be **lower** than a designated flight level and not likely to infringe a control area/airway.

ABSENCE OF A PRESSURE SETTING FACILITY

Certain types of small, usually recreational, aircraft may be flown using an altimeter such as a wrist type which may not have a pressure setting facility. The answer in the UK, where altimeters are usually set to show 0 feet on landing, lies in compromise by adopting the following procedure:
- Prior to take-off adjust the altimeter to read the height in feet of your departure point. At least, having made the adjustment, you will be flying cross-country as near as you can to QNH – the current sea level pressure.
- Just prior to arrival overhead your destination, *reduce* your altimeter reading by the destination height above mean sea level (amsl) as shown on your aeronautical chart.
- Pressure setting will now be QFE – the airfield pressure.
- On landing the altimeter should read 0 ft.

For example:

- Departure point 550 ft amsl – Destination 160 ft amsl.
- High ground to 800 ft *en route* – flight to be at 1800 ft.
- Altimeter set at 550 ft prior to take-off.
- After take-off there will be a 1250 ft climb to 1800 ft above msl.
- Just prior to arrival – *reduce* altimeter reading by 160 ft.
- Altimeter now reads 1640 ft – your height above the destination airfield.
- Pressure setting will now be QFE.
- On landing the altimeter reading should be 0 ft.

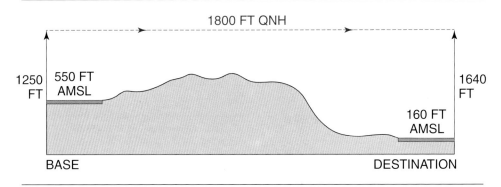

To arrive at a destination with a QFE setting, take off on QNH and above the destination reduce the altimeter by the height of the airfield amsl

You must understand that this compromise does not allow for an *en route* QNH (or atmospheric pressure) setting change. However, the type of aircraft likely to be flown in these circumstances is unlikely to be one that is airborne for any length of time or be capable of covering a distance long enough to make any presssure change too much of a problem.

ALTIMETER SETTINGS – THE UNITED STATES

In the USA the scene is rather different. there is no 'Q' code and no airfield pressure setting such as QFE found in the UK. Also, there is no variable transition altitude – it is fixed at 18,000 feet. Above this level altimeter settings are at the ISA standard of 29.92 in.Hg (1013 mb).

Altitude in flight below transition altitude is measured against mean sea level pressure (msl) (UK QNH) which is simply referred to as the altimeter setting at the current atmospheric pressure (msl) of the nearest station within 100 nautical miles (nm).

Thus, the altimeter set at msl pressure will not only show the height of the airfield above mean sea level as you take off – it will also show the destination airfield's height on landing. Every airfield has its height amsl clearly shown on charts.

Throughout the USA, unlike the UK, you will be fortunate to have constant personal contact with a met centre whilst in the air in order to secure the latest MSL pressure setting within 100 miles at any stage of your flight.

PRACTICE SHEET (FOR UK PILOTS)

See page 192 for an illustrated Altitude Scale

1

To find pressure altitude related to a given QNH altitude

1. Using the Pressure Altitude scale, measure the QNH altitude and mark on a ruler.
2. Set 0 feet of the measurement against the current QNH value on the QNH Altitude scale.
3. Mark off the QNH altitude with an 'x' on the QNH scale.
4. Place ruler level with 'x' on QNH scale to Pressure Altitude scale.
5. Read off the pressure altitude equivalent to the QNH altitude on the Pressure Altitude scale.

2

To establish transition level after changing to SPS at the Transition altitude

1. Establish pressure altitude related to QNH altitude as per 1 to 5 in the previous exercise above.
2. Transition level is the first 500 foot level above the established pressure altitude on the Pressure Altitude scale.

NOTE:

The above exercises will show near enough accurate answers but the real aim of the exercise is for you to understand the principle.

CHAPTER FIVE

WIND

DIRECTION AND SPEED

Wind is simply the movement of air from one place to another. It is reported in terms of direction/speed referred to as a wind velocity (W/V).

The reported direction (W) is that *from* which it is blowing; stated in compass degrees from 001° to 360° related to True North.

The speed (V) is reported in **knots** (nautical miles per hour). Other units of wind speed used can be miles per hour (MPH), kilometres per hour (KPH) and metres per second (MPS). We will confine ourselves to the officially accepted unit of knots.

Thus, a South-Westerly wind of 15 knots would be reported as a W/V of 225/15.

For aviation purposes the direction is to the nearest 10 degrees, so the above wind reported to a pilot would be 230/15. As just stated, a wind direction is related to true north on all occasions – but there is one exception.

Runways being related to magnetic north lead to this direction being used in wind reports from a control tower. Therefore, with a variation of 8°W, a wind from 270 True would be 262° Magnetic – reported as 260° (the nearest ten).

The instruments used for

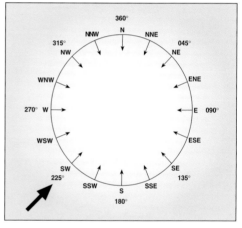

A south-westerly wind

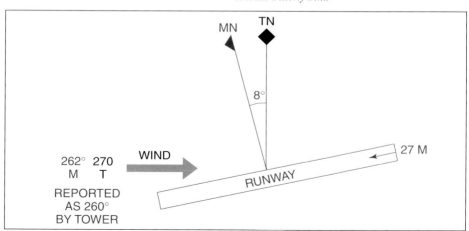

Wind direction related to magnetic north

The anemometer

establishing wind direction and measuring speed are known as the **wind vane** and **anemometer** respectively. They are placed at a minimum of 10 metres (30 ft) above the surface and located to be well clear of turbulence at ground level. The current wind direction/speed is then transferred to become a visual presentation within the met centre.

Changes in wind direction are known as **veering** when clockwise and **backing** when anti-clockwise. For example, in a change from say West (270°) to South-West (225°), the wind is said to have backed; when the change is from say North (360°) to North-East (045°), it is said to have veered.

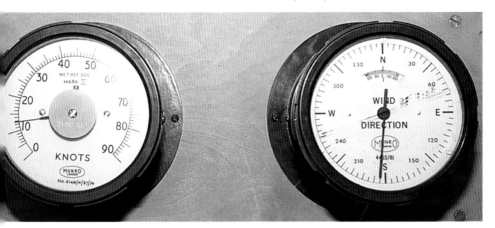

The speed/direction indicator

CAUSE OF WIND

So how does wind come about? Imagine two containers – one full of water and the other empty. Pipe them together at the base and both will end up holding the same amounts of liquid. The water has achieved a balanced state called **equilibrium**.

Wind direction changes – clockwise is veering – anti-clockwise is backing

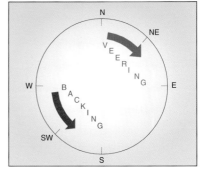

Air behaves in the same manner as water; it seeks equilibrium and so flows from an area of high pressure to one of low pressure. But it does so in rather a roundabout way.

The force that seeks to move the air direct from high to low is called the **pressure gradient force,** which in future we shall refer to as simply the **gradient force**. The gradient is defined as **steep** or **shallow** according to the distance between isobars – similar to contours on a map where the closer they are, the steeper is the incline.

Just as a stream will be fast moving down a steep slope, where the map

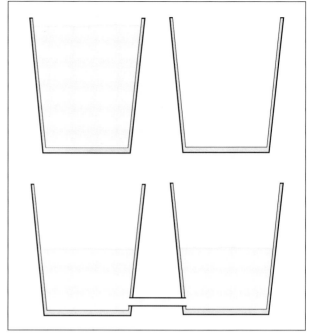

The water will flow until equilibrium is reached

contours are close together, so wind speed is directly related to the gradient, being strong when steep and lighter when shallow – or **slack**, as it is sometimes called.

However, if the gradient force had its own way there would be a direct rush from high to low and the result would be alarming! Quite simply this does not happen; observation alone will show you that changes in wind speed are normally gradual with a largely constant speed being maintained over many hours – sometimes days. So what slows down the potential rush?

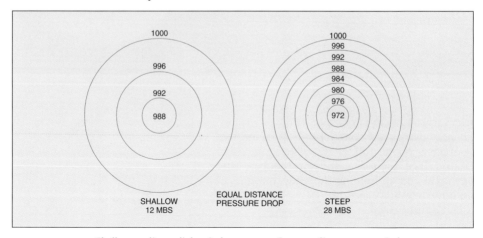

Shallow gradient – light winds   *Steep gradient – strong winds*

RELATION OF WINDS TO PRESSURE SYSTEMS

The gradient force is in fact balanced by another 'force' – known as **Coriolis Effect**. This effect, not strictly speaking a force, is caused by the Earth's rotation. It is greatest in the high latitudes and fades out at the Equator.

Its effect results in the air, except at the Equator, moving along or parallel to the isobars and not across them. The direction will be as follows.

Clockwise	around a **high** (anticyclone)
Anti-clockwise	around a **low** (depression)

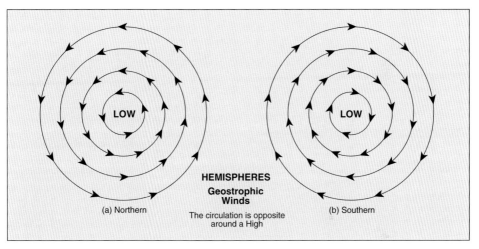

Geostrophic winds around a low-pressure centre

Note that in the Southern Hemisphere these directions are reversed, being clockwise around a low and anti-clockwise around a high.

CORIOLIS EFFECT

Coriolis Effect is called an 'effect' because it is not really a force in its own right. Imagine a ball-bearing set in motion 'downhill' from the North Pole towards the Equator – but not in contact with the ground. Now view this ball-bearing from point A on the Equator immediately below its start point at the North Pole.

As the ball moves down, the observer will be moving to the right with the Earth's rotation so at A1 the ball will appear to have moved to the left or clockwise to the right from the ball's viewpoint. Continue this exercise by noting its position as seen from A2, A3 and A4.

Note that when the ball is moving from south to north, its path is now to the left or anti-clockwise. Although the illustration suggests the ball is coming from the South Pole it does not necessarily have to be in the Southern Hemisphere to do so. A northerly path will always be anti-clockwise. Due to the Earth's rotation.

Coriolis Effect is the 'force' which appears to 'pull' the object into its circular path. Remember, it is greatest in the Polar regions and fades away at the Equator.

Now replace the ball with air which is on the move. Coriolis Effect pulls the flow into a circular motion around a pressure system and holds it there; but the gradient force wants it to go towards the low pressure centre. The flow is maintained along the isobars because Coriolis Effect and gradient force balance each other out. Any increase in the gradient force will result in an increase in Coriolis Effect with an overall increase in the speed of the flow.

It was mentioned earlier that Coriolis Effect was minimal in Equatorial regions. This means that the pressure gradient has a greater influence and the flow of air takes place more directly from high to low pressure. The pattern of this flow is indicated by an **isotach** *– a line joining places of equal wind speed.*

The lines so formed are known as **streamlines***. As pressure gradients in the tropical regions are not as great as those in the higher altitudes, the result of the tendency for a direct flow from high to low pressure is not dramatic.*

To the observer at point A the ball will appear to be due north. As the observer moves to point B due to the Earth's rotation the ball will appear to be moving away to the observer's left and will continue to do so until finally it is completely to his left. From the ball's point of view relative to the Earth's surface it has been deflecting to the right in relation to its north-south path

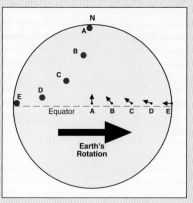
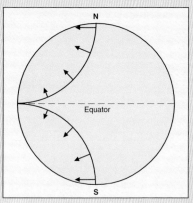

The Coriolis Effect – notice how it fades out at the Equator

Coriolis Effect holds good around 2000 ft and above; the flow along the isobars is known as a **geostrophic wind**. It is measured by means of a **geostrophic scale** placed across the isobars on a chart; this measurement plays a major part in forecasting. Shortly you will see that a true geostrophic wind is pretty rare but the isobar/geostrophic scale measurement approach is sufficiently accurate to give the forecaster a good working base.

Buys Ballot's Law states that when you stand with your back to the wind, the low pressure area is to your left in the Northern Hemisphere and to your right in the Southern Hemisphere.

Since winds blow *around* pressure centres you may well be thinking 'If the air always does this, how can it ever move from high to low to establish equilibrium?' A good question.

Friction effect

Take a pack of playing cards and release it with a swinging motion over a card table. The bottom card will immediately stop on contact with the cloth surface whilst the other cards will spread themselves out along the line of motion until the top card is well ahead of the bottom one. Friction on contact with the table has slowed down the bottom cards.

So it is with an airflow (wind) below around 2000 ft. It is slowed down by the friction of the Earth's surface and this is increased by the presence of mountains, forests, towns and so forth. As the wind speed slows down, the Coriolis Effect, which is normally in balance with the gradient force, is weakened but, in this case the gradient force is allowed to exert itself.

This results in the surface wind moving inwards towards a low and outwards from a high at an angle of around 25 to 30 degrees over land. With less friction over sea the change is only around 10 degrees.

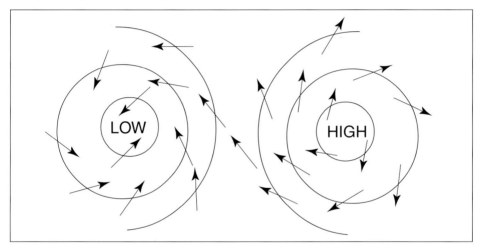

The airflow from high to low pressure

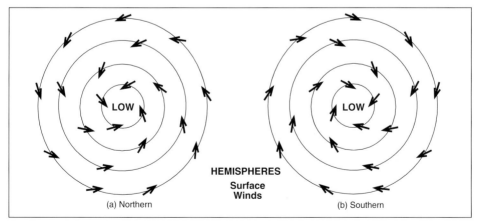

The effect of friction on wind direction at the Earth's surface can be seen in the deflection towards the centre of the low

A good way of seeing friction effect for yourself is to note the direction of the surface wind and compare it with the direction of cloud movement at around 2000 ft. But remember, the surface change in direction means an adjustment to Buys Ballot's Law.

The speed of the wind at the surface over land decreases to approximately 33% of that at the geostrophic wind level, and 66% of same over the sea. In case you may think it is a golden rule that windspeed continues to increase with height above 2,000 ft, this is not so. Whether it increases with height or not will depend purely on the pressure gradient existing at a given level at the time. Very strong winds can develop just under the tropopause but this is not directly due to increasing height.

Converging isobar effect

A true geostrophic wind only exists when the isobars are straight and parallel to each other, with no disturbance to the balance between the gradient force and Coriolis Effect. However, isobars will invariably be on the move either closing up (converging) or moving apart (diverging). When converging, the gradient force can *temporarily* override the Coriolis Effect and, once again, the flow deviates towards the low pressure centre.

> *When parallel isobars converge, the gradient force increases. Before the Coriolis Effect can catch up to be in balance with it, the gradient force moves the flow to the left towards the low pressure centre (or right in the Southern Hemisphere).*
>
> *When the change in pressure gradient is very rapid the deflection inwards towards the low can be both sharp and as much as 45 degrees.*
>
> *This process will continue until the isobars are once again parallel, the airflow having adjusted to a constant speed where Coriolis Effect is once again in balance with the gradient force.*

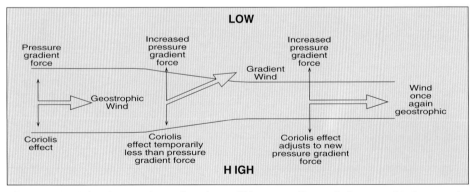

The converging isobars increase the gradient force, which temporarily exceeds the geostrophic force, causing the wind direction to back and flow towards the low-pressure centre

GRADIENT WIND

When the flow is diverted from the geostrophic path towards the low pressure it is known in the UK as the **gradient wind**; in the USA it can be known as the **resultant wind**.

WIND SPEEDS AROUND PRESSURE SYSTEMS

Winds can blow faster around an anticyclone than they do around a depression.

An analogy best explains why this happens. Take an object and whirl it round your head at the end of a piece of string. As you let go, the object flies off at a tangent to the circle it formed whilst rotating.

The force causing the object to fly off is known as centrifugal force and such a force exists with air blowing around a high pressure system.

Moving around a high, centrifugal force wants to make the air escape from its circular path. This force in fact becomes an addition to the gradient force which in turn causes the Coriolis Effect to increase as they both seek to balance each other. The result is an overall increase in wind speed around the system.

Conversely, the outward centri- fugal force around a low has to compete with the inward gradient force and this reduces the overall gradient force and thus slows down the flow. In both the above cases it is assumed that the gradient force is constant at the time.

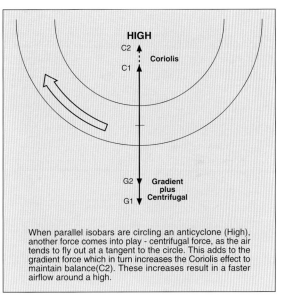

When parallel isobars are circling an anticyclone (High), another force comes into play - centrifugal force, as the air tends to fly out at a tangent to the circle. This adds to the gradient force which in turn increases the Coriolis effect to maintain balance(C2). These increases result in a faster airflow around a high.

The outward centrifugal force increases the pressure gradient force (G) from G1 to G2 and the Coriolis Effect (C) will then move from C1 to C2

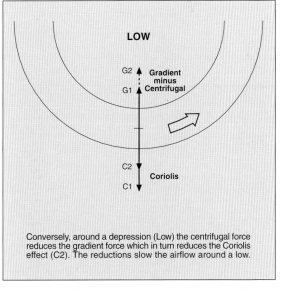

Conversely, around a depression (Low) the centrifugal force reduces the gradient force which in turn reduces the Coriolis effect (C2). The reductions slow the airflow around a low.

Here the centrifugal force acting in the opposite direction reduces the pressure gradient force from G1 to G2 and the Coriolis Effect responds by reducing from C1 to C2

DIURNAL (DAILY) VARIATION

Wind speeds can vary with the times of day being faster during the day and easing off at night – given that no change in the existing pressure system is taking place.

When the sun is up and the atmosphere in proximity to the surface is buoyant, the difference between the surface and faster geostrophic wind aloft is reduced. At night when the sun is not around the difference can be at its maximum, with the surface wind even dropping to nil but the geostrophic wind remaining the same. The 'buoyancy' referred to is to do with the stability or instability in the atmosphere and will be dealt with later.

THE JET STREAM

There can be a very fast upper wind known as a **jet stream**. This flow may be moving at 90 to 200 knots and resembles a form of ribbon of considerable length but relatively little width and depth. Winds surrounding a jet stream's core are far less vigorous.

VERTICAL MOTION

Although not related to a wind as such, at this juncture mention should be made of vertical movement in the atmosphere where a region of air can be imperceptibly ascending or descending. It is relatively fractional compared with the horizontal flow of wind in terms of speed but it plays a major part in the development of depressions (lows) and of cloud formation.

This vertical motion should not be confused with the vertical motion caused by convection, turbulence or fronts, which have yet to be discussed.

CHAPTER SIX

EFFECTS OF WIND ON FLIGHT

LOW-LEVEL TURBULENCE

The atmosphere is rarely at rest – particularly near to the ground. The surface of a deep river looks placid yet it can be hiding considerable turbulence taking place on the bed.

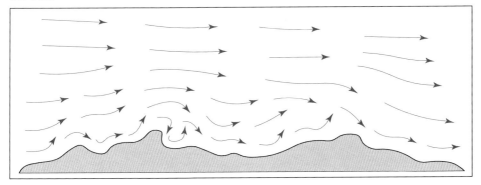

At low level, think of how a flow of water would be over a rough surface

If the river is shallow the behaviour of the water close to the bed is more apparent; ripples, waves and eddies caused by rocks and ruts are visible to the eye. It has already been said that air behaves like a fluid, so picture the atmosphere as being this river but with the rocks and ruts on the bed now being trees, buildings, towns, or mountain ranges on the surface of the Earth.

The waves and eddies in the air will be invisible but their effect on an aircraft close to the ground can be potentially very dangerous – particularly in the case of light and very light aircraft.

Always have a mental picture of what the air can be doing when you are flying at low level with obstructions in the vicinity.

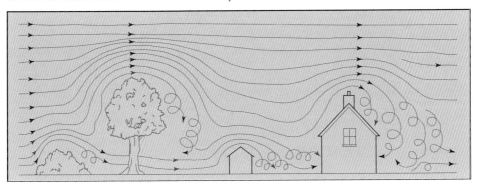

An airflow behaves in exactly the same way as the flow of water on a rough river bed

MOUNTAIN/HILL EFFECT

Be ever conscious of your position in relation to mountains and hills. On the leeward (downwind) side you may be cruising close to that side at the correct engine revs and airspeed, only to find yourself descending at an unhealthy rate towards the ground.

An instinctive reaction at that moment may be to climb; this could result in a stall where power is insufficient to cope with the increased angle of attack induced by your pitch-up input. There is only one proper course of action – increase your airspeed and head downwind away from the hills to escape the downdraught.

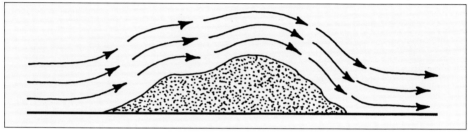

Descending air to the lee of a hill or mountain range

WAVE EFFECT

Still staying with our river analogy, an obstacle can set up a wave which does not cease on passing the obstacle but can continue to oscillate and form a series of waves for some considerable distance downstream.

A mountain or hill range can create the same effect with air. In this situation, when flying parallel to the range, you can experience the unwanted descent referred to previously; but this time it can be many miles downwind from those hills. This is simply because you are in the air descending on the downwind (lee) side of the crest of one of these waves.

To avoid this belt of descending air you should turn upwind or downwind of your existing track and then resume your original heading when you are out of the

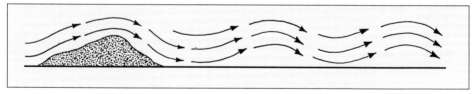

Descending air in wave effect can have the same effect as would be experienced to the lee of a mountain

effect – flying either along the trough or on the crest of the wave!

In fact play your cards right and detect the *ascending* air on the upwind (windward) side of the wave. Here you will find lift which could well allow you to continue at reduced power without any drop in speed – a good fuel saver! This assumes your route continues to be parallel to the hills creating the wave effect.

Such waves can reach up to great heights – particularly in the United States – and are much sought after by sailplane pilots, as you will see later when we discuss lenticular clouds.

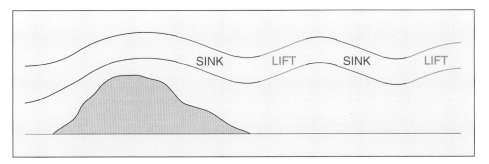

The areas of sink and lift related to a lee wave

Lenticular cloud at top left produced from a wave. Notice the undulation also affecting a layer of cloud below

HILL LIFT

Hills and mountains can be a pilot's friend – ask any sailplane pilot. Hill lift is the vertical component of an airflow being forced up a slope.

In (a) you are looking at the obvious.

In (b) you are looking at a breakdown of the forces involved.

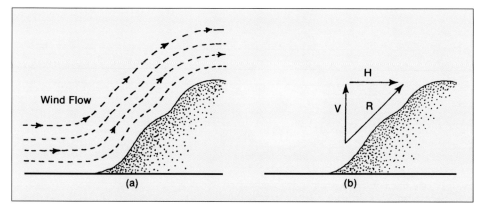

R – The resultant flow of air due to hill.
H – The horizontal component of the resultant.
L – The vertical component of the resultant – lift.

For example, a wind speed of approximately 20 mph – say 1800 ft per minute (fpm) – could produce a vertical component of some 1200 fpm in a given area above the slope.

Thus a sailplane descending at say 600 fpm through the air rising at 1200 fpm would result in it ascending at a rate of 600 fpm (10 ft per second) in relation to the ground as long as it remained in the rising air.

The area of best hill lift is to be found about one-third out from the top of the hill in relation to the foot of the hill.

There have been occasions when pilots in light aeroplanes have experienced power problems and have moved into available hill lift to extend their flight – even reaching their destination by doing so. Such moves have prevented a precautionary or forced landings.

ROTOR EFFECT

A serious situation can be encountered on the lee side of a mountain range which is very steep. The descending air can curl back under itself to form a large closed eddy, called a **rotor**, with turbulent cloud perhaps forming at the top of it and clearly seen to be rotating.

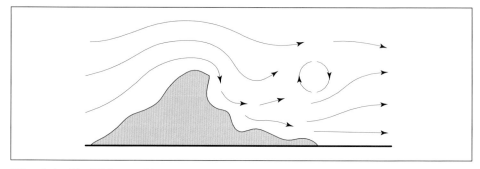

When the lee side of high ground is steep, rotor effect can exist in strong winds

Rotor cloud to the left of hills - well worth avoiding

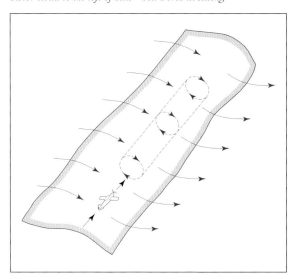

Note how rotor could affect the angle of attack on your wing as you enter it

You will appreciate that the chance of a light aeroplane's angle of attack being where it should be is remote in the changing airflow in a rotor, even if such a machine has the power to escape its clutches.

A good example of rotor in the UK is to be found in Cumbria when a strong cold north-easterly wind (known as a **Helm Wind**) blows over the top of the Cross Fell range in late winter and spring and plunges down the western slope. It is usually located above a point from just over ½ to just under 4 miles (1 to 6 km) from the foot of the slope.

The windward side of the mountain must not be ignored when the sloping side sharply flattens out at the top. Just as the airflow breaks down over a wing when the angle of attack is too high, so the air flow over the top of a steep mountain or hill with a sharp crest will do likewise. This potential danger needs careful consideration when landing after slope soaring.

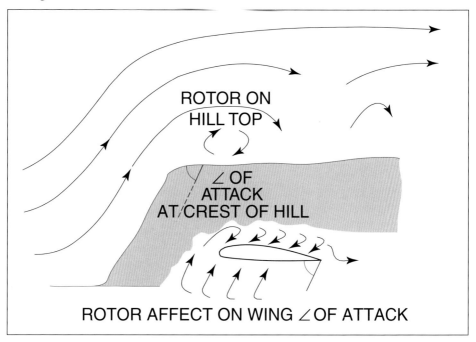

Compare the rotor effect on the windward slope with a very high angle of attack on a wing

MOUNTAIN/VALLEY EFFECT ON WIND DIRECTION

A surface wind crossing a mountain range can be suddenly diverted from its path when faced with a valley going off at an angle to the wind flow. Should the valley be narrow there will not only be a change in direction but a substantial increase in turbulence as well as speed.

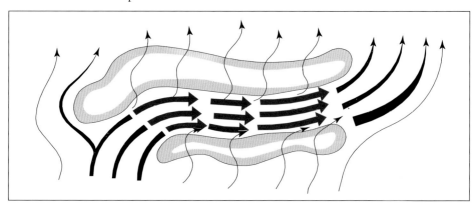

Not only can winds speed up on being confined within a valley, they can also create severe turbulence

WIND SHEAR

Wind shear in meteorological terms broadly refers to a change in both direction and speed of wind. It is described as **vertical wind shear** when related to changes with height and **horizontal wind shear** when the changes take place at the same level.

Vertical wind shear

Over many years the change of wind speed with height (vertical wind shear) in *immediate proximity to the surface* has often been referred to as **wind gradient** (not to be confused with pressure gradient or gradient wind) and its effect known as **gradient effect**. This is mentioned so that in dialogue with other (possibly older!) pilots you will know what the terms imply.

We discussed earlier the behaviour of the pack of cards and this analogy highlighted the drop in windspeed near the surface due to friction. The reverse of course happens on climb-out when there is a sudden increase in windspeed with height.

On take-off you will climb into faster-moving air away from the surface and show a *temporary* increase in airspeed. Do not be tempted to reduce your airspeed; if you do, in all probability a moment later, due to the effect wearing off, you will find yourself back to the original pre-reduction speed. The danger is that you may find yourself at a *lower* figure than you had originally because you had made too great a reduction; if that speed is below stalling speed you could have a problem relatively close to the ground.

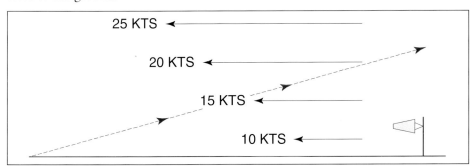

On take-off a temporary increase in airspeed can occur: do not compensate for this increase

Conversely, on landing the effect can be even more critical as you are fast running out of height. It is more applicable to the slower-moving type of aircraft where the wind speed can be a marked proportion of the approach speed. As you enter the zone of reduced wind speed near the ground, your airspeed can *temporarily* drop away and your aircraft suddenly sink. If your descent has been made at an approach speed just above stalling speed you are likely to faced with an expensive repair bill.

Although not specifically associated with meteorology, the same effect of sudden sink could see you making an even bigger mistake by initiating a pitch-up input and inducing a stall, leading to the same repair bill as you end 'up to your wings in airfield'!

The answer lies in an adequate approach speed, say an addition of one-third (33%) to one-half (50%) of the windspeed, based on the fact that the stronger the wind, the more marked will be the gradient effect – or vertical wind shear in modern parlance.

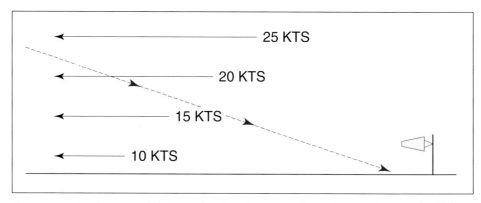

On an approach to land a marked decrease in airspeed can occur close to the ground; prepare for this by an increased approach speed equivalent to between a third and a half of the windspeed

Again, turning too steeply in close proximity to the ground can pose a potential problem. In a turn the outer wing travels faster than the inner wing, resulting in an increased (outer) and decreased (inner) airspeed over each wing respectively.

In the bank associated with the turn the outer wing will be higher than the inner wing and due to gradient effect can be subject to a further momentary airspeed increase on the higher (outer) wing and decrease on the lower (inner) wing.

This applies particularly to aircraft with a large wingspan such as sailplanes.

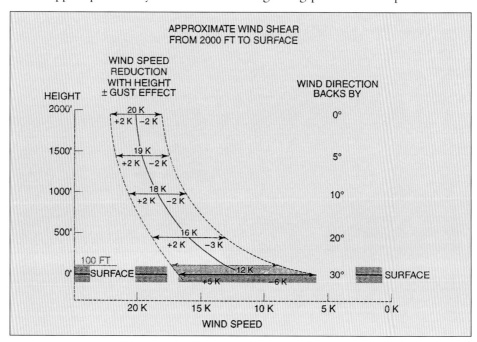

A very rough idea of the variations that can take place with vertical wind shear. Particularly note the change close to the surface

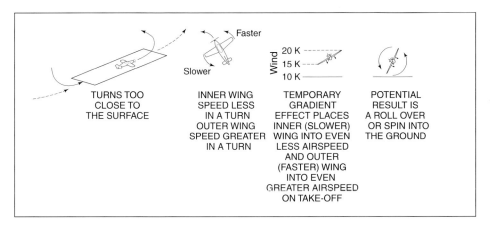

| TURNS TOO CLOSE TO THE SURFACE | INNER WING SPEED LESS IN A TURN OUTER WING SPEED GREATER IN A TURN | TEMPORARY GRADIENT EFFECT PLACES INNER (SLOWER) WING INTO EVEN LESS AIRSPEED AND OUTER (FASTER) WING INTO EVEN GREATER AIRSPEED ON TAKE-OFF | POTENTIAL RESULT IS A ROLL OVER OR SPIN INTO THE GROUND |

Horizontal wind shear

On a conventional landing approach into wind you have just seen how vertical wind shear (or gradient effect) can cause a sudden drop in airspeed. The same effect can be produced by a change in wind direction on your approach, as any such change will bring about a reduction in the value of your head-wind component.

The main cause will normally be the presence of a cumulonimbus cloud in the vicinity. The reason *why* will come later when discussing convection and this cloud type in detail. In the meantime, take my word for it; it can cause a change in wind direction and bring about a horizontal wind shear effect as well as a vertical effect.

At (A) We have the normal pattern of events.

      (B) Horizontal wind shear has taken place; related to the aircraft the wind is now part crosswind and part headwind.

      (C) The headwind component is now half the total windspeed in this example.

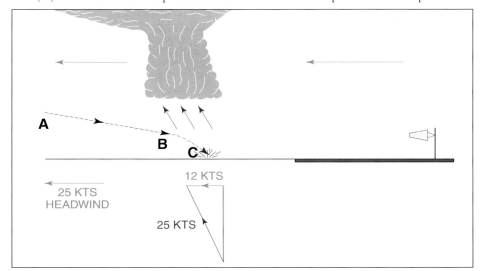

A change in wind direction can reduce your headwind component, resulting in a decrease in airspeed to stall point

Usually when conditions are right for localised horizontal wind shear of any magnitude, the weather is potentially poor with thunderstorms in the vicinity – certainly not the conditions in which small aircraft should airborne.

The effect on the aircraft can be the same as that for vertical wind shear except it can sometimes be more hazardous – particularly if the change in direction is from a point to the rear of the aircraft, when it suddenly becomes a tailwind.

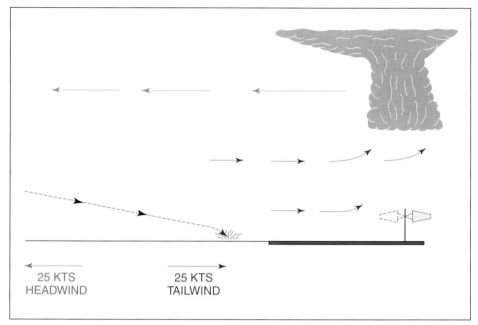

This situation can have an adverse affect on all sizes of aircraft

GUSTS

Wind is rarely constant in speed in view of the barriers in its path, as shown in discussing low-level turbulence. Gusts are composed of sudden increases and decreases in speed which are an added hazard in landing through wind gradient (vertical shear).

SQUALL

Similar to a gust but more prolonged in time. It is very much associated with cumulonimbus cloud and, again, will be dealt with later.

WINDS FROM YOUR LEFT
(OR RIGHT IN THE SOUTHERN HEMISPHERE)

When you experience a strong wind coming at you from your left-hand side, could also be identified by a drift to the right, it could mean that you are heading towards a depression (Buys Ballot's Law).

Repetition here will do no harm. The lower pressure into which you are heading will cause your altimeter to read an increased altitude when in fact you are not climbing.

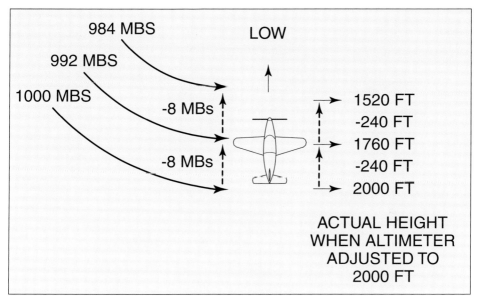

In this situation, if not adjusted the altimeter will eventually read 2480 ft. If repeatedly corrected en route to read 2000 ft it will result in an actual altitude of 1520 ft

OTHER WINDS

Wind can also be associated with temperature changes, but discussion on this aspect will be left until temperature itself has been explained.

WIND CHILL FACTOR

When the air is calm the heat of the human body warms the air in immediate contact with the skin, forming a kind of 'protective' layer. As a wind develops this layer is eroded and in very cold regions such as the Arctic or Antarctic it can be dangerous in the effect it can have on the exposed parts of the body.

Those of you flying aircraft in which you can be subject to exposure should be aware of wind chill as the heights to which an aircraft can reach could well match apparent Arctic temperatures.

WAKE TURBULENCE

Although not specifically related to this chapter, every type of aeroplane will produce wake turbulence, and wind plays a part in its path and therefore has an effect.

The width of vortex effect is related to the wing span of the aircraft from whose wing tips the majority of turbulence emanates as a spiral.

It is at its maximum when the aircraft is flying slowly at a high angle of attack, which is largely the case on take-off and landing. A crosswind can play a part in moving the wake to one side or the other of a runway.

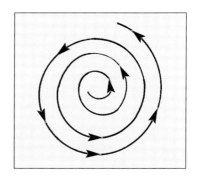

A cross-section of the airflow in wake turbulence

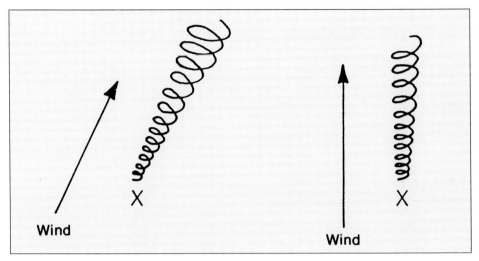

Wind influences the path of turbulence on leaving a wing

Likewise, buoyancy in the atmosphere will possibly see it rise above the flight path of the aircraft following behind. You cannot therefore guarantee that it will only exist directly behind the aircraft being followed; in flight it could be above or below the flight path.

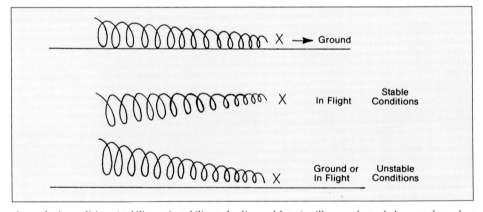

Atmospheric conditions (stability or instability to be discussed later) will cause the turbulence to descend or ascend accordingly

Vortices usually last for up to 80 seconds but this can increase to around two and a half minutes when the ambient air is dead still and very dense in cold conditions.

Do bear in mind that these vortices can roll an aircraft over with such force that the ailerons have no effect in the matter. In case you might think the phenomenon is confined to large aircraft only then beware; it can happen to one microlight behind another and has done – I have experienced it and was not at all happy!

Helicopters can produce wake turbulence exceeding that of a large aircraft.

The CAA produce a booklet on wake turbulence but at present there seems to be no figures specific to the light aircraft weight category and below.

CHAPTER SEVEN

TEMPERATURE

ORIGIN

Temperature is simply the measured warmth or lack of it in the Earth's atmosphere. Perhaps contrary to popular belief, the atmosphere does not obtain all its heat directly from the sun. The sun heats the Earth by a process called **insolation** and the Earth gives off the heat gained by a process called **radiation**. In other words the Sun can be considered as a giant boiler with the Earth being a very large radiator.

The farther north or south you go from the Equator the more is temperature influenced by the seasons. For example, in the British Isles during summer the sun is high in the sky and the days are long, with the opposite occurring in the winter.

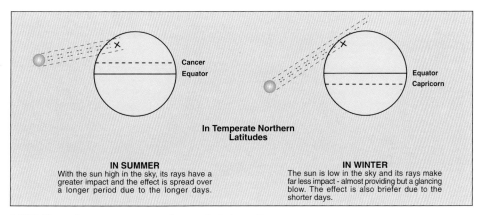

In Temperate Northern Latitudes

IN SUMMER
With the sun high in the sky, its rays have a greater impact and the effect is spread over a longer period due to the longer days.

IN WINTER
The sun is low in the sky and its rays make far less impact - almost providing but a glancing blow. The effect is also briefer due to the shorter days.

LEFT: The sun's rays are concentrated on a relatively small area
RIGHT: The sun's rays are spread over a larger area and for less time in winter

Cloud acts as an insulator; it can reduce the amount of heat taken in by the Earth during daylight and also reduce the amount of heat escaping into space through radiation at night.

MEASUREMENT

In the UK, and many other countries, temperature is currently measured in degrees **Celsius (C)** (originally called Centigrade). Except when dealing with upper air levels, in the USA it is still reported in **Fahrenheit (F)** – a scale used in the UK until some years ago.

The difference is as follows:	Celsius	Fahrenheit
Freezing point	0°C	32°F
Boiling point	100°C	212°F

Conversions:

Celsius to Fahrenheit	Fahrenheit to Celsius
Deg $\dfrac{(C \times 9)}{5} + 32$	Deg $\dfrac{(F-32) \times 5}{9}$

The **ISA Standard** at MSL is **15° Celsius (59° Fahrenheit)**.

The temperature at any given location, height and time in the atmosphere is known as the **ambient** temperature.

RELATIONSHIP OF TEMPERATURE WITH HEIGHT

Generally temperature will decrease with height. If you think it should increase because you are getting nearer the sun then try climbing the Rockies in a swim-suit! In fact it's a proof of the Earth heating the atmosphere because climbing away from the surface takes you farther from the radiator.

The atmosphere enclosing the Earth is made up of an inner 'layer' called the **Troposphere** and an outer 'layer' called the **Stratosphere**. The dividing line between the two is called the **Tropopause**.

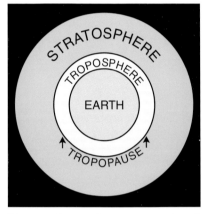

Within the Troposphere a steady decrease in temperature with height normally takes place. The **ISA Standard** for temperature decrease with height is set at **1.98°C per 1000 ft (300 metres)** but accepted figures of **2°C** or **3.6°F** per 1000 ft are normally used.

From this you can roughly assess the temperature at a given height. With a surface temperature of 20°C (68°F) it could be said that at 5000 ft the temperature could be 10°C (50°F). The words 'could be' are used because the decrease per 1000 ft of 2°C (3.6°F) is a theoretical standard which in practice constantly varies.

In the lower levels of the Stratosphere the temperature remains virtually constant at an **ISA** figure of around **–55°C**, after which comes a slight increase in its upper levels.

The troposphere reaches up to approximately 11 km (35,000 feet) above the surface in the temperate latitudes

There are other regions above the Stratosphere but this book is not for budding astronauts so we will ignore them.

Temperature governs the height to which the Troposphere ascends above the surface of the Earth, so it varies with its geographical location and with the seasons. It is highest over the Equator and lowest over the Poles. It will also be higher over a region in summer than in winter.

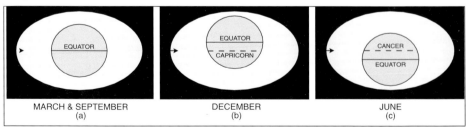

MARCH & SEPTEMBER (a)	DECEMBER (b)	JUNE (c)
March and September over the Equator	*December over Capricorn*	*June over Cancer*

HIGHEST POINT OF TROPOSPHERE

Flying above the winter weather in the stratosphere

At 42,000 feet the curvature of the Earth is just becoming apparent

It is within the Troposphere that virtually all weather and cloud formation occur. Hence the saying 'flying above the weather' when passing through the Stratosphere.

TEMPERATURE EFFECT ON DENSITY

As touched upon earlier, when a parcel of air of a given volume is heated, the air expands and fewer molecules then occupy the space of the original given volume. From this you can deduce that

an increase in temperature　　means　　**a decrease in density**.

Conversely,

a decrease in temperature　　means　　**an increase in density**.

UNEVEN SURFACE HEATING

The easiest way to introduce this topic is by asking you a question. On a hot summer day, when out for a picnic or a barbecue, would you sit on the grass or an outcrop of rock? There are no prizes for the answer; suffice it to say the surface heat of the rock would be uncomfortable whereas the surface heat of the grass would be acceptable.

In the interests of simplicity we will not go into any complex explanations concerning specific heat, as it is called. Broadly speaking, the solid nature of rock precludes a deep penetration so all the heat received is confined to the surface.

At the other extreme, water is easily penetrated and, therefore, the same amount of heat that beat down upon the rock will penetrate water and spread itself out with little or no particular concentration at the surface. In fact the sea can take a whole summer season to warm up to its maximum temperature.

Here is a selection of surfaces indicating roughly their order in terms of the variation in absorption related to them.

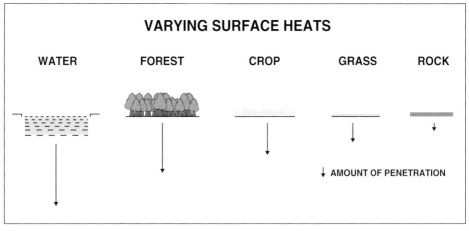

The variations shown are related to the 'specific heat' of each substance – a term used in physics which is a subject you were promised would be avoided

The important point about uneven surface heating is the fact that air immediately in contact with the surface will become warm or cool according to the temperature of that surface. When it becomes warmer and less dense than the surrounding air the surface air will seek to rise. Two further analogies will prove the point.

Consider a garden bonfire. As it burns you will see sparks and hot embers shooting into the air. They are caught up in the rising hot air thus providing a visual example of hot air rising, and the fiercer the heat the faster they will shoot skywards.

Similarly, think of the hot air balloon and the periodic bursts of flame from its burner. The air within the balloon is being heated to a greater temperature than the ambient or environment temperature, which causes it to ascend.

INVERSION

There are times when the temperature will *increase* with height; such an occurrence is called an **inversion**

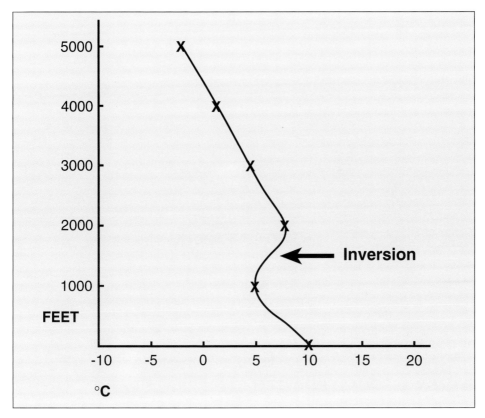

An inversion

Take the evening; the sun has set and, given a clear sky with no insulating blanket of cloud, there can be a very rapid loss of heat at the surface in a short space of time. The air in contact with the surface can become colder than the air above which has been warmed by the heat given off during radiation. This can produce a well-defined inversion by the next morning.

A sure sign of an inversion forming can be seen at any time when smoke from a chimney, in calm or next to no wind, immediately flattens out as it tries to ascend.

A frequent sight in the early evening

Such inversions usually disappear soon after the sun has risen and the Earth's surface has once again warmed up. Dispersal can also come about when a wind arises; the ensuing turbulence lifts and mixes the colder air with the warmer air thus negating the temperature difference between the two. The effects of an inversion on flight are dealt with later.

Inversions are not confined to surface levels; they can occur at any height.

WINDS ASSOCIATED WITH TEMPERATURE

Sea breeze effect

During the day the land heats up much more quickly than the sea, causing the air over the land to become less dense. Being less dense and therefore lighter than the air over the sea it will rise.

When established at its new level the risen air plus the existing air at that level brings about an increase in pressure level. The now higher pressure over the land, when related to the pressure over the sea, initiates a gradient force with a subsequent flow of air out to sea at that level. In other words a circulatory pattern is set up.

The surface air now flows in from the sea to replace the gap left by the rising air over the land, usually at a speed of around 10 knots. It is this flow which is known as a **sea breeze**.

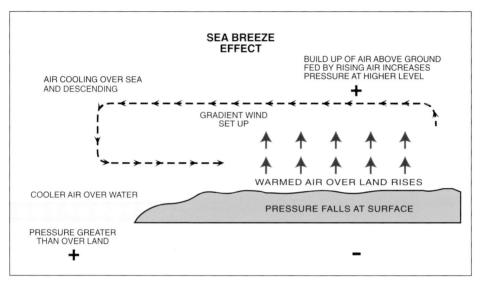

The effect of a sea breeze can reach some way inland

Given that an off-shore wind (blowing from the land) is not too great, the sea breeze can over-ride it and penetrate some way inland. Gradually, as the day goes on, due to Coriolis Effect it can tend to veer and blow obliquely or even parallel to the coastline instead of directly towards it.

Care should be taken when you arrive overhead to land at a coastal airfield. When descending from circuit height, you have to consider the nature of both vertical and horizontal wind shear that can be caused by the differing direction/speed of any prevailing wind blowing out to sea when related to the sea breeze direction/speed.

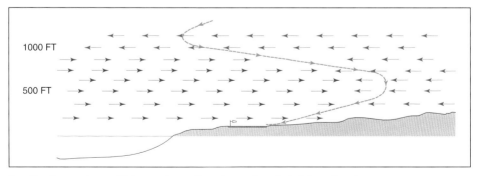

Landing in a sea breeze effect can take you through a variety of wind shear situations

This will be apart from the usual effect of wind gradient (vertical shear) on the last stages of approach.

Ensure you check the windsock if there is no controller to give you a current wind report.

'Convection' wind

'Convection' wind is not an official term. I use it to describe that momentary wind caused when air warmed by uneven surface heating breaks away from the ground. As it rises, surface air then flows in from all sides to replace it. This occurrence can often be seen on a calm, hot summer day when you feel a slight breeze or see a windsock suddenly lift for a moment or two before once again becoming limp. Later we shall discuss convection as a process in its own right.

The gap left by a thermal breaking away from the surface can create a temporary wind which flows in to fill the gap

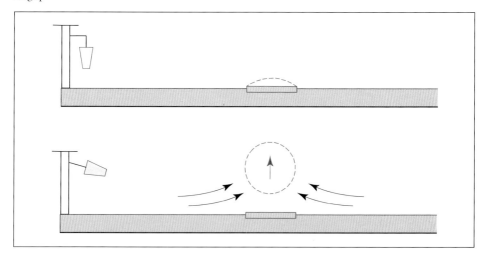

A lazy windsock momentarily comes to life as a thermal breaks away from the surface and ascends

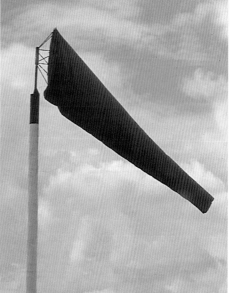

Katabatic wind

This is a wind which can be caused by air cooled at the top of mountains or hills as the sun goes down in the evening. As it cools the air becomes denser and heavier and flows down the slopes into the valley or plain below. It can be relatively strong, with gravity assisting its descent.

To save a walk down the hill in late evening, hang-glider pilots have been known to attempt taking off to fly to the bottom and have then wondered why they never became airborne – they had tried to take off downwind. The term **katabatic** in relation to descending air will crop up again later.

On a larger scale, picture the Rockies where air forced to rise to around 12,000 ft can cool and end up so cold compared with its surround that it cascades down to the plains below at a very considerable speed. In this location it is called the Chinook; in the European Alps it is known as a Föhn wind.

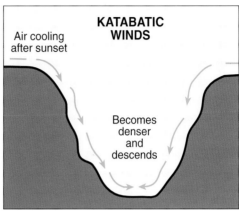

The cold air can sometimes form a pool in the valley where in winter frost can form while the surrounding area remains clear of it. Such a location is known as a frost hollow

Anabatic wind

This is the reverse of a katabatic wind. When the early morning sun warms the lower slopes of a valley or mountain; the air in contact with them also warms and rises up the slopes. An anabatic wind will be weaker than the katabatic wind as it is opposed by gravity in its ascent. The term **anabatic** in relation to ascending air will also crop up again later.

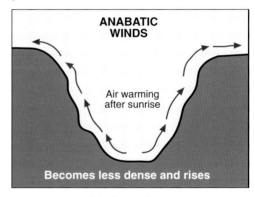

An anabatic wind in the morning

Wind chill temperatures

When discussing 'Wind Chill' earlier on it was said that the apparent temperature could decrease to polar levels in a strong wind. Here is an example of a less exciting but still unpleasant situation: it is calm and the temperature is as low as 2°C (36°F). A wind starts to blow and on reaching around 20 knots it will produce an apparent temperature of around –12°C (10°F). This is a drop of 14°C (26°F).

Other aspects of temperature

There are other aspects associated with temperature such as upper winds, sea breeze fronts and turbulence, which will unfold as we progress to areas where they more readily fit in and you have acquired further knowledge.

CHAPTER EIGHT

HUMIDITY

Humidity is concerned with the moisture content of the atmosphere, ever present in the form of invisible water vapour in varying amounts.

There is a limit to the amount of water vapour that a parcel or given mass of air can hold. To quantify humidity, the amount of vapour present in a mass of air at a given time is related to the total amount of vapour that the mass is capable of containing. The water vapour content is then expressed as a percentage of that total and is known as **relative humidity**.

Temperature has a direct bearing on relative humidity; the warmer the air the greater the amount of vapour that it can hold. So, on warming, the same amount of water vapour present in the air would now be *less* when related to the increased potential total. In other words, the relative humidity of the air will have decreased.

Conversely, on cooling, the air can only contain less water vapour. Therefore any vapour present at the time the cooling takes place will now be *greater* when related to the total potential and the relative humidity would have increased.

If the cooling of moist air continues, the point is reached where the air can no longer contain any further water vapour – the relative humidity has reached 100%. The air is now described as **saturated air** and the temperature at which this occurs is called the **dew point**.

Should still further cooling below the dew point take place the excess water vapour over the 100% will change into visible water particles which is cloud.

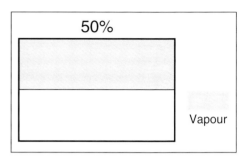

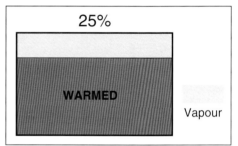

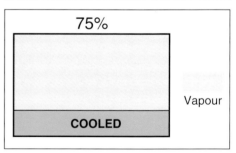

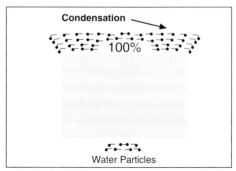

The process is known as **condensation** and for this to occur minute particles of matter have to be present in the atmosphere upon which condensation can take place. This matter is known as **condensation nuclei**.

Condensation becoming visible as cloud

In the case of cloud, the height at which condensation takes place is called the **condensation level** which will broadly coincide with the **dew point level** if the nuclei are present.

On the rare occasions when such nuclei are not present below dew point and the relative humidity exceeds 100%, the air still cooling is said to be **super-saturated air**.

When temperatures drop below freezing point at 0°C (32°F) the logical result should be the water vapour in excess of 100% relative humidity changing into ice. However, for this to happen there must be other types of nuclei present in the atmosphere – known as **freezing nuclei**.

The process of ice changing directly from water vapour into ice crystals is known in aviation meteorological terms as **sublimation**. The term is also used to describe the transformation of ice crystals directly back into water vapour.

*For those who are 'into' science, the term sublimation may be disputed in respect of the change of vapour into ice. The process of vapour turning directly into a solid is known in other scientific quarters as **deposition**. Sublimation is the change of a solid into a vapour. However, we must stick with the term used in the world of aviation meteorology to avoid confusion.*

In the absence of freezing nuclei an alternative transformation is the water droplets striking a frozen surface, whereupon they will change into ice.

Until either of the above factors is introduced the water droplets will remain in liquid state in the form of **super-cooled droplets**, and can do so even down to as low as −45°C (−49°F).

Here are the various processes and their affect on temperature:

Vapour to Liquid	– Condensation	– Warming
Vapour into Solid	– Sublimation (Deposition)	– Warming
Liquid into Solid	– Freezing	– Warming
Solid into Liquid	– Melting	– Cooling
Solid into Vapour	– Sublimation	– Cooling
Liquid into Vapour	– Evaporation	– Cooling
Summing up: Relative humidity	– **decreases with warming**	
and	– **increases with cooling**.	

Also, remember there is a relationship between relative humidity and density because water vapour is less dense (lighter) than dry air.

- **high humidity – lower density**
- **low humidity – greater density**

CHAPTER NINE

EFFECT ON FLIGHT OF THE BASIC FACTORS IN THE ATMOSPHERE

Having now covered the basic factors of density, pressure, temperature and humidity it is time to see the important part they all play in an aircraft's performance. For interest's sake the following are very approximate picture of the atmosphere relatively close to the surface based on the ISA mean sea level standard.

Height	Pressure		Temperature		Density
feet	mb	in.Hg	C	F	
10,000	700	20.60	–5°C	23°F	74%
5000	850	25.10	5°C	41°F	87%
2500	930	27.43	10°C	50°F	93%
SURFACE MSL	**1013.25 mb (hPa)**	**29.92 in.Hg**	**15°C**	**59°F**	**100% 1.225 kg per cubic metre**

You may now begin to think you are diverting from meteorology. Not so; the issue concerns conditions which are specifically related to flight safety – a key element in the aim of this book.

EFFECT ON LIFT

Earlier on, under 'Density' it was said that the calculated performance of an aeroplane was based on the ISA (International Standard Atmosphere) of 1013.25 mb (29.92 in.Hg) and 15° Celsius (59° Fahrenheit) producing a standard density of 100% – all at MSL (Mean Sea Level). Where take-off performance is concerned, to this should be added one non-meteorological factor; the assessment of take-off run assumes using hard ground.

You now need to recall and remember; when the pressure is *lower* than the ISA figure of 1013.25 mb (29.92 in. Hg) and/or the temperature is *higher* than the ISA 15°C (59°F), density decreases and so does the performance of the aeroplane. You will require a longer take-off run and the rate of climb will be less.

As an example we will assume that a take-off is being made at mean sea level with a pressure setting of 1013 mb (29.92 in.Hg), a temperature of 15°C (59°F) and a 100% density of 1.225 kg per cubic metre. In other words the conditions match those of the ISA standard. We will also assume a nil wind.

Lift at take-off requires a given airspeed. Your aircraft requires, say, an airspeed of 72 knots to become airborne so at a ground speed of 72 knots in nil wind it should lift off the runway.

The picture would be thus.

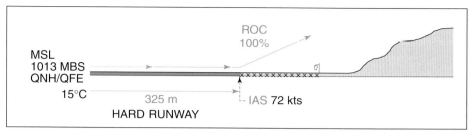

Take-off under ISA conditions

Now take a simple grass strip airfield with the following criteria. It is at 500 ft above mean sea level (amsl) with the QNH at the time being 993 mb (29.32 in.Hg) and it is a hot summer day at 25°C (77°F). The field is 500 metres (1640 ft) long and you normally use 325 metres (1066 ft) to get airborne.

Under these conditions, at the 72 knots ground speed, the less dense (thinner) air will mean less dynamic pressure (the pressure caused by the speed of air into the pitot tube). This leads to an airspeed reading less than the actual speed along the ground – giving an indicated airspeed (IAS) of around 69 knots. As it is sufficient *airspeed that counts to become airborne* you are now faced with a longer take-off run because you have to build up a faster ground speed to produce the required airspeed of 72 kts.

It can be calculated that, under the above conditions, take-off would require an extra 33% of runway amounting to an additional 107 metres (351 ft), making a total of 432 metres (1417 ft) to reach lift off. On the strip in question you would now be faced with a margin of 68 metres instead of 125 metres or 223 ft, instead of 410 ft in which to get airborne. Again, I suggest you follow through with your own diagrams to scale to see for yourself.

Such a take-off would in any case be deemed unwise in view of the need to clear possible obstacles. But wait; the calculations against the ISA standard are based on a take off from *hard ground* and the strip in question is grass! In actual fact, a take-off from grass as opposed to a hard surface requires a further 20% (30% if wet grass!) additional distance to lift-off, which in this example would be 432 metres + 20% or just over 518 metres (1700 ft) on your 500 metre (1640 ft) strip!

If you attempted a take-off in these conditions it would surely see you 'up to the cockpit in hedge' unless you hopefully 'sussed' something was wrong early on in the take-off run and called off the attempt.

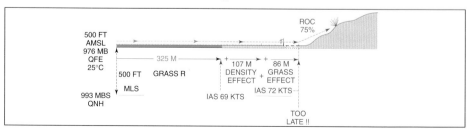

Take-off under non-ISA conditions

The effect does not stop at take-off. The climb rate in the above case is also affected by the less dense air to the tune of around 25% *less* than normal for the given conditions in this example. Would you be able to clear any obstructions ahead?

If you feel that all this is 'stretching it a bit' – visit Denver in Colorado which is 5000 ft above sea level; here the additional take-off distance required can be 70% plus. Leaving there one day in a Boeing 727, the take-off run seemed so long I was convinced we were going to Florida by road!

Finally, remember that the **ground speed** on landing will also be faster in situations such as those elaborated above.

Note, all the above figures of effect are approximate. Your aeroplane manual should provide data specific to your machine.

As a broad yardstick here is a table of approximates to provide a useful guide. In the USA, where there is no QFE, the pressure altitude would be used by setting 29.92 in.Hg on the altimeter.

		Airfield Temperatures								
Pressure	*ISA*	*Standard*			*Non-Standard*					
Altitude	*QFE*	*15°C or 59°F*			*20°C or 68°F*			*25°C or 77°F*		
feet	*mb*	*Dist*	*%*	*ROC*	*Dist*	*%*	*ROC*	*Dist*	*%*	*ROC*
0	1013	0		0	+ 7		− 5	11		− 10
400	1000	+ 6		− 4	+ 12		− 10	+ 20		− 17
1150	975	+ 16		− 6	+ 24		− 20	+ 33		− 25
1900	950	+ 30		− 22	+ 35		− 25	+ 42		− 30
2650	925	+ 40		− 30	+ 50		− 33	+ 59		− 38
3400	900	+ 54		− 36	+ 63		− 41	+ 75		− 46
4900	850	+ 93		− 53	+ 102		− 56	+ 120		− 60
KEY:	Dist	is % extra distance required for take-off								
	ROC	is % reduction in rate of climb.								

You must understand that the QFE figures and related heights in the above table assume the pressure to be the ISA standard of 1013 mb and the airfield to be at mean sea level – a combination rarely found in reality. Two further adjustments are needed where pressure and temperature are not the standard figures, and yet again it is suggested you make your own diagrams to help you understand and grasp the process.

1. If the actual QNH is, say, 993 mb then at a sea level airfield this would also be the QFE and would be the 'Airfield QFE' figure to use in the table.

    or

2. When your airfield is above mean sea level you must reduce the ISA sea level 'Airfield QFE' of 1013 mb in the table by the number of millibars that equate to the height of the airfield.

    Eg: The airfield is 500 ft amsl. At 1 mb per 30 ft this is a drop of 17 mb. The 'Airfield QFE' would now be 996 mb (1013 – 17).

Combining the two conditions given above would mean the sea level 'Airfield QFE' would be 993 mb (1013 – 20), being the difference from ISA, and a further –17 mb for airfield height amsl, giving a figure of 976 mb (1013 – 37 in total) to be used when consulting the table.

It also means that related to the ISA standard, the 'pressure altitude' based on the QFE is not 500 but 1100 ft (1013 – 976 mb x 30 ft). In the USA it is the 1100 ft pressure altitude you would use – **not** the 500 ft height above MSL figure.

EFFECT ON POWER

You may have wondered why, apart from the reduced density calling for an increase in ground speed to reach flying speed, it takes so long to reach the required flying speed when the difference in the example is only 3 knots. It is because less density also results in a decreased power output.

As density decreases, the ratio of air to fuel in a carburettor will change. The air input will be less making the fuel input relatively greater, thus leading to a loss of power due to the over-rich mixture. Similarly, the thinner air produces less 'bite' from the propeller and less dense is the air expelled from a jet.

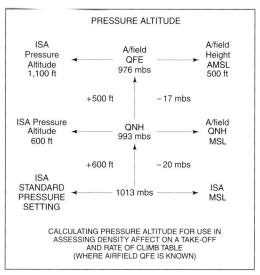

Calculating density altitude in the UK and the USA; there is no QFE setting in America

Upper Air Readings

It is not envisaged that the readers of this book will be flying in upper air regions at this stage. So, from an interest point only, the effect of a pressure/density/temperature decrease is most marked in the upper air. This greatly reduces static pressure input into the altimeter and dynamic pressure into your airspeed indicator (ASI).

Calculations to assess this effect on your altimeter and ASI readings in order to obtain true readings can be made on your navigational computer – it is assumed you would have one in your flight bag.

HUMIDITY

Whilst air with a high relative humidity will be less dense than dry air, the effect on flight is relatively minimal except perhaps in tropical conditions.

THE FINAL WARNING

Do please remember; with an airfield way above sea level and/or given a hot humid day, you must seriously consider the effect of all these density related factors *before* attempting to take off from a small field – particularly where obstructions or high ground have to be avoided during a reduced climb rate.

Not to do so poses the possibility, I say again, of ending 'up to your cockpit in hedge' (hopefully not trees) – a painful experience with damage to the wallet if not the body!

CHAPTER TEN

LAPSE RATES

A lapse rate is the rate at which temperature decreases with height. You will recall the accepted figure is 2°C (3.6° F) per 1000 ft. However, this is but an average; there are other types of lapse rate we must discuss.

ENVIRONMENTAL LAPSE RATE (ELR)

The **environmental lapse rate (ELR)** is the rate at which air cools with height in the atmosphere at a given time; in other words, changes in ambient temperature. It is a variable as it is constantly changing; hence the need for readings to be taken at frequent intervals.

ADIABATIC LAPSE RATE (ALR)

This is the lapse rate which applies to any parcel or mass of air being *forced* to rise up through the atmosphere. On rising it is subjected to less pressure as it gains height and cools **adiabatically**. For such rising air to cool adiabatically it must not be subject to any loss or gain of heat from an exterior source.

The best way to explain **adiabatic cooling** is to use another analogy and then reverse the findings! Take a bicycle pump; after a short period of use it becomes hot without any additional heat being applied from any exterior source. The increased pressure applied by the piston in the cylinder has produced the heat.

Conversely, when pressure is *decreased*, cooling takes place; think of the way in which a refrigerator works.

When adiabatic cooling takes place it does so at a *fixed* rate known as the **adiabatic lapse rate** of which there are two forms – the

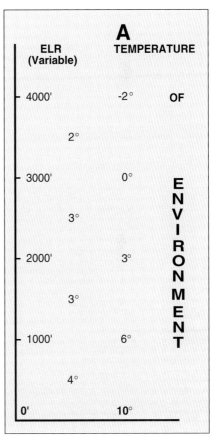

Column A depicts the temperature of the atmosphere at each 300 metre (1000 foot) level according to the environmental lapse rate (ELR)

Dry Adiabatic Lapse Rate (DALR)
and the
Saturated Adiabatic Lapse Rate (SALR).

Dry Adiabatic Lapse Rate (DALR)

This lapse rate applies to rising air which has not condensed. It is at a *fixed* rate of **3°C (5.4°F)** per 1000 ft.

Saturated Adiabatic Lapse Rate (SALR)

This lapse rate is also sometimes referred to as the 'wet' adiabatic lapse rate and it takes over from the DALR as soon as condensation takes place.

During the process of condensation heat is given off; it is called **latent heat** and once again requires an analogy to explain it in simple terms. Apply a hair dryer to your head and you will notice a cold feeling until your hair is dry. The process of evaporation uses up heat. Conversely, the opposite process of condensation gives off heat.

Latent heat continues to be given off as long as condensation is taking place and it reduces the DALR by half to the SALR – now a *fixed* rate at **1.5°C (2.7°F)** per 1000 ft.

> *This rate remains good until above approximately 6000 ft. It then begins to increase until at around 33,000 ft there is little or no difference between the SALR and the DALR. This is because the upper air is so cold that the water vapour content is minimal and therefore the amount of potential latent heat will also be minimal.*

Adiabatic lapse rates, when compared to the environ-mental lapse rate, indicate the

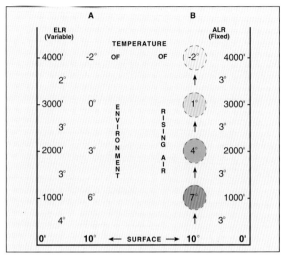

Column A depicts the temperature of the atmosphere at each 300 metre (1000 foot) level according to the environmental lapse rate (ELR). Column B depicts the temperature of rising air within the atmosphere which cools on ascent at the fixed adiabatic lapse rate (ALR) of 3°C

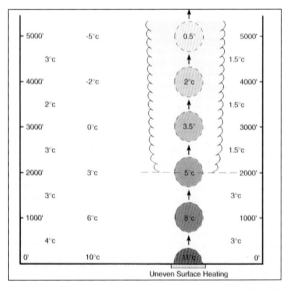

On reaching dew point the change to a saturated adiabatic lapse rate can substantially increase the warmth within the cloud and accelerate the ascent

condition of the atmosphere at any given time. They also have a direct bearing on cloud formation.

STABILITY

When the **ELR** *is less than the ALR* ascending air can find itself cooler and thus more dense (heavier) than its environment. It therefore ceases to rise and will tend to sink back towards the surface. In such a case the atmosphere is said to be **stable**.

INSTABILITY

When the **ELR** *is greater than the ALR*, ascending air will be warmer than its environment and thus less dense (lighter). It will therefore continue to ascend, and in this case the atmosphere is said to be **unstable**. The rising air will only cease to rise when its temperature has become the same as that of its environment, when it has reached **equilibrium**.

Stable and unstable layers of air can exist at any level. For example, the air at the surface could be unstable and become stable a few thousand feet higher and vice versa.

So far we have talked about adiabatic cooling. It should be noted that there is also **adiabatic warming** – recall the bicycle pump analogy. Here *descending* air heats up at the DALR.

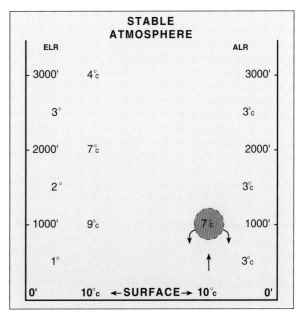

The rising air would be 2°C cooler than its surround at 300 metres (1000 feet) and therefore being cooler (heavier) it would tend to sink towards the surface. In fact the rising air would cease its ascent before reaching the 300 metres (1000 feet) depicted; the ascent would stop as it reached the same temperature as the environment – in this case around 80 metres (250 feet)

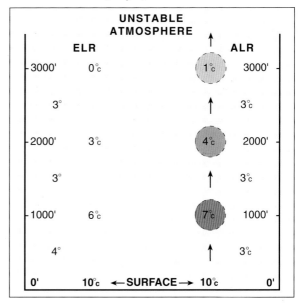

As long as the rising air is warmer than its surround, it will continue to rise until equilibrium is reached

Here is an example of how air can be warmer on the leeward side at the foot of a mountain range than it would be at the foot of the windward side.

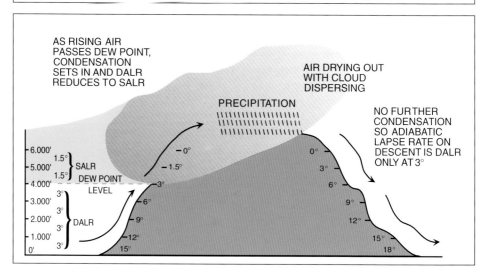

AS RISING AIR
PASSES DEW POINT,
CONDENSATION
SETS IN AND DALR
REDUCES TO SALR

AIR DRYING OUT
WITH CLOUD
DISPERSING

PRECIPITATION

NO FURTHER
CONDENSATION
SO ADIABATIC
LAPSE RATE ON
DESCENT IS DALR
ONLY AT 3°

With the cloud having deposited most of its moisture, the air becomes less moist and the relative humidity falls below 100%, allowing evaporation to take place, thus thinning and dispersing the cloud. With no further condensation taking place, the saturated adiabatic lapse rate (SALR) changes back to the dry rate (DALR) for the descending air

CHAPTER ELEVEN

ORIGIN OF CLOUDS

Cloud is formed when moist air forced to ascend in the atmosphere cools and condenses into visible water particles. There are five ways in which such an ascent can happen.

Orographic cloud

Orographic

When air comes up against a mountain range some of it may manage to go round the sides but the main mass will have no alternative but to climb over the top. When a geographical feature causes air to rise the process is known as orographic.

Turbulence

Earlier we saw how turbulence can be set up at low level by the presence of obstructions on the Earth's surface. The vertical element in any turbulence aloft can lift warm moist air to below dew point and result in condensation.

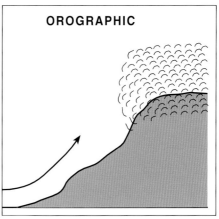

OROGRAPHIC

The ascending air will be cooling at an adiabatic lapse rate

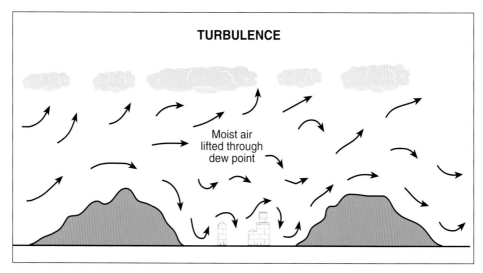

Surface obstructions distort a smooth airflow causing turbulence

Convergence

This process is set in motion when air over a region rises from the surface to replace air removing itself from the upper levels over that region. This situation will be discussed in more detail later under 'Depressions'.

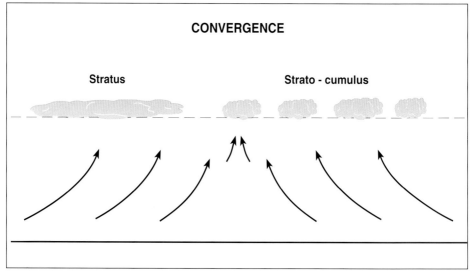

The height at which cloud may form will depend on the dew point level at the time and the presence of condensation nuclei

Frontal

A mass of warm air meeting a mass of cold air, being less dense, will not be able to penetrate the colder denser mass and therefore it climbs over it. The difference between warm and cold air is relative; a parcel or layer of air at 25°C (77°F) will be defined as 'cold' in relation to neighbouring air at 30°C (86°F). Likewise, –10°C (14°F) is 'warm' related to air at –15°C (5°F).

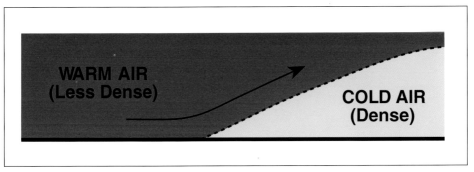

The frontal ascent is similar to the orographic situation, with the colder, denser air acting as an invisible mountain range

Convection

This process is set in motion by a parcel of air ascending after being warmed to a greater temperature than its environment.

Refer back to the uneven heating of the Earth's surface discussed earlier: surfaces liable to high surface heat, such as towns, rock outcrops, runways, can be surrounded by, perhaps, woods or fields, the surface heat of which is lower because of greater absorption on their part.

Where a surface is warmer than its surround, the air immediately in contact with it will also become warmer and, given an unstable atmosphere (ELR > ALR), the air will break away from the surface and ascend. The process that triggers off this ascent is known as **convection** and was touched upon earlier under 'Winds Associated with Temperature' and 'Lapse Rates'.

Convection can also be triggered off when warm air is lifted into a colder unstable layer of air – for example, by orographic or frontal means.

THE THERMAL

The parcel or bubble of air rising from a surface due to convection is known as a **thermal.** Thermals, along with hill lift, are the very essence of the sailplane's ability to remain airborne and, indeed, soar to extraordinary heights albeit in cloud at times.

For those who wish to soar, the thermal is great news; the sailplane pilot is letting his aircraft go with the flow to gain height. However, keeping in the thermal and maximising its lift is another story and there are plenty of books on the techniques required to do so.

To the light aeroplane *en route* thermals are looked upon as turbulent air where, suddenly, one is rising, and on coming out of them is the feeling of a sudden descent – often referred to as an 'air pocket' by the uninformed.

Strong thermals can in fact be 'worked' by very light aeroplanes as well as by sailplanes, hang gliders and para-gliders. But back to the matter in hand. The lesson you learn from all the above is that in unstable air you can experience a rough passage, the extent of which will depend on the strength of convection at the time.

Large cumulus clouds resulting from strong thermal activity (convection) prevalent at the time took this sailplane to 14,000 feet

ESTIMATING CLOUD BASE HEIGHT

Given that you know the height of the dew point level, (and assuming condensation nuclei will be present), you can reasonably say that at this height cloud will begin to form. For example:

Surface temperature is	15° C	59° F
Dew point temperature is	6° C	43° F
Diff'ce in surface to dew point is	9° C	16° F
DALR of rising per 1000 feet	3° C	5.4°F
Est. Height of cloud base will be:		
Temp. diff'ce ÷ DALR x 1000 ft =	9 ÷ 3 or	16° ÷ 5.4°
	x1000 ft	
	= 3 x 1000	3 x 1000
Estimated base will be 3000 ft		

Here is a recap of the five ways in which air may be forced to rise:

•	Climbing over mountains	–	OROGRAPHIC
•	Mixing at a given level	–	TURBULENCE
•	Pressure adjustment	–	CONVERGENCE
•	Air masses meeting	–	FRONTAL
•	Uneven surface heating (or by entering unstable air, eg orographic climb)	–	CONVECTION

We must now begin our journey through the clouds – pictorially as well as by means of the written text.

CHAPTER TWELVE

CLOUDS

CLASSIFICATION

A cloud can fall into one of three basic classifications which can be

HIGH	MEDIUM	LOW

according to the height of its base. This will vary for each classification according to its location on the Earth and the season of the year.

Recall what was said earlier about the Troposphere where weather and cloud exist: the height or top of this belt (the Tropopause) was at a maximum over the hot Equator and at its minimum over the cold Poles.

In the temperate latitudes wherein exist the British Isles, the seasonal effect is quite substantial and the *approximate* base for each classification can be as follows:

HIGH	Summer	24,000 ft	(7500 m)
	Winter	16,500 ft	(5000 m)
MEDIUM	Summer	16,500 ft	(5000 m)
	Winter	6500 ft	(2000 m)
LOW	Summer	6500 ft	(2000 m)
	Winter	1000 ft	(300 m)

To go to extremes, high cloud in the tropical summer can be found up to 45,000 ft (14,500 m). In a polar winter it could be as low as 10,000 ft (3000 m).

CATEGORIES

Within the three classifications there are two categories:

Category:	**Stratiform (Str)**	or	**Cumuliform (Cu)**
Basic Format:	Layer Cloud		Heap Cloud
Conditions:	Stable Atmosphere		Unstable Atmosphere

Mist and fog are in a sense cloud, but they will be dealt with separately at a later stage.

Apart from the types of cloud within each classification and category, there can be a number of variants within each type; the only ones included in this book, however, are those which are pertinent to you as a pilot.

We will take high cloud first and work our way down.

HIGH CLOUDS

High cloud in isolation has one feature which distinguishes it from any other; it is composed of ice crystals. There are three main types.

Cirrus (ci)

White fibrous hair-like wisps or strands which are virtually always transparent. Sometimes the strands can be turned up at one end (giving rise to the description 'mares' tails'), caused by ice crystals falling away from the source point and trailing back in an area of less wind lower down.

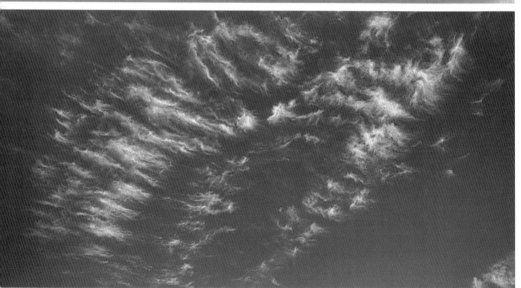

Cirrus (Ci)

Cirro-cumulus (Cc)

Relatively thin white patches or sheets of globules sometimes formed into ripples known as a 'mackerel sky'.

Cirro-cumulus (Cc)

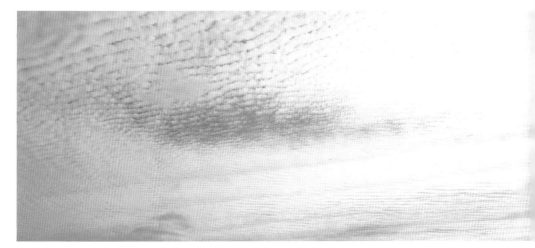

A truly mackerel sky!

Cirro-stratus (Cs)

Cirro-stratus (Cs): a thin veil or sheet of white fibrous cloud which can sometimes cover the whole sky

Cirro-stratus from above

The sun will shine through cirro-stratus, and occasionally a halo will be seen around the sun.

A solar halo

MEDIUM CLOUDS

Medium cloud can closely resemble cirrus except that its colour may be darker. It is also not so fibrous being composed mainly of water droplets. There can however, be an ice crystal content when the cloud is very high.

The three main types come under the Stratiform (layer) category.

Alto-cumulus (Ac)

White or grey or a mixture of both. Of globular structure similar to cirro-cu but, being at a lower level, the globules appear larger.

Alto-cumulus (Ac)

An alto-cumulus layer seen below just before climbing up through cirro-stratus

The 'crazy paving' effect of alto-cumulus

Alto-cumulus
Lenticularis (lent)

Lenticular clouds can appear at the crest of the waves generated by mountain ranges as discussed earlier under 'The Effect of Wind'. Ideally, for such clouds to form the wind must be fairly strong and be blowing directly on to the mountains. If the wind direction is too oblique the effect will diminish. We have already touched upon the lift available to sailplane pilots on the windward side of a wave; lenticulars are a very useful indication of such conditions.

They can also serve as a warning to pilots of light aeroplanes that they may experience that sudden rate of descent mentioned earlier, where airspeed and power are nevertheless showing normal.

Unless there is sufficient moisture in the air, lenticular clouds may not necessarily appear, or at least not at the level at which you are flying, but the effect could still be there in the clear air downwind of the mountains.

Alto-cu lenticularis (Ac Lent): cigar shaped cloud forming at the crest of a standing wave

A lenticular at the top of a wave wearing a close-fitting cap!

The trough between two waves

Alto-stratus (As): greyish sheet or sheets of cloud, thicker than cirro-st and not so fibrous

Alto-stratus (As)

Greyish sheets or one single sheet which can cover the whole sky. Thicker than cirro-stratus and not so fibrous. The sun may shine through thin alto-stratus but it would be hazy in appearance.

There are two types of medium cloud in the cumuliform category which are closely associated with thundery conditions; they will be covered when we discuss 'Thunderstorms' later on.

The next type of medium cloud, although originating at medium level, should really be classed as low cloud because it is most apparent at this level. Another example of the grey areas in meteorology!

Thin alto-stratus allows the sun to show through — albeit opaque in appearance

Nimbo-stratus (Ns)

A thickening of alto-stratus which eventually extends down to low cloud level. It is amorphous in appearance with any shape or base being obliterated by the heavy rain it produces. It is invariably associated with frontal systems.

Nimbo-stratus (Ns): an extension of thick alto-stratus whose base ultimately extends down to low cloud level

The shape of nimbo-stratus is virtually impossible to define due to obliteration by associated rain

LOW CLOUDS

Strato-cumulus (Sc) and Stratiform (Str)

Similar to alto-cumulus but appearing larger. Lumpy, perhaps white on occasions, but mostly greyish in colour. The lumps or rolls can be independent or merged together with an undulating surface.

Strato-cumulus (Sc): separate lumps with blue sky in between

Strato-cumulus (Sc): merged together and can be distinguished from stratus by the undulating under surface viewed from the ground and the air

Stratus (St)

Similar to alto-stratus but usually at very low level. It will be greyish in colour and devoid of any specific shape.

Very much associated with hills and lifting fog.

Stratus (St) can appear as one sheet covering the whole sky or as several sheets. It will be greyish in colour and devoid of any form or shape

Fracto-stratus (Fs)

Fragments of stratus usually seen when stratus is in the process of forming or decaying.

Fracto-stratus

The breaking up of a stratus layer

Fracto-nimbus (Fn)

Fragments of cloud, appearing similar to fracto-stratus, which form under nimbo-stratus when air saturated by rain is lifted in turbulence and condenses. Sometimes referred to as 'scud' cloud.

CUMULIFORM (Cu)

Probably the most well known category of cloud. Although the tops of well developed cumuliform cloud can reach up *above* the troposphere on occasions, it is the normal height of the base that governs the classification of cumulus as low cloud.

There are many variations within this category and it should be added that a spin-off from developed cumulus can be medium and high clouds.

Remember, cu cloud primarily stems from convection due to uneven surface heating, or an orographic climb into an unstable atmosphere. The larger cu is very much associated with turbulence, ranging from moderate to horrific!

Fracto-cumulus (Fc)

Can vary from a wisp to a lump of ragged edged cloud – white in colour. It will not have the flat-bottomed base of the ordinary small cu. Given the conditions at the time it will either disappear or grow into a conventional cu cloud.

Fracto-cumulus (Cu fra): Taken from above amid growing small cumulus

Small cumulus (Cu)

Easily identifiable by its flat base and rounded top. It is designated small when the base is greater than the height. It is sometimes referred to as 'fair weather cu' when seen on a summer day.

Small cumulus interspread with Cu fra

Small Cu from above growing larger

Large cumulus (Cu)

A larger form of cu cloud where the height exceeds the width of the base. This type often develops into a towering mass with the well-known 'cauliflower' tops. This distinct type of large cu is called 'congestus'.

Large cumulus

Traditional 'cauliflower' cloud with new growth on the right

Developing cumulus soon to become part of its cu-nimb companions on the left

Cumulus genitus (Cugen)

A developing cu may reach equilibrium at any stage of its development on coming up against a stable layer whereupon it ceases to climb. When this occurs it can spread out as stratiform cloud at low or medium level in the form either of strato- or alto-cumulus according to the height at which the stable layer is reached.

Cumulus cloud spreading out into genitus cloud on reaching equilibrium

This layer cloud will continue to spread out as long as convection lower down continues to feed it. Such cloud is known as alto-cumulus-cumulus-genitus(!) or cugen for short. The name cugen denotes that it was generated by cu cloud and did not form of its own volition. This type is the 'enemy' of sailplane pilots as it can blot out the sun and substantially reduce the convection generated from uneven surface heating.

Cumulonimbus (Cb)

The most powerful of all clouds – its appearance will be depicted in its stages of growth and decay.

1. Cumulonimbus calvus (Cb cal): the tops begin to lose their clear-cut 'cauliflower' definition and merge into a smoother top of 'bald' appearance

2. The next stage can be fibrous wisps spreading out from the top.

3. On reaching the tropopause, or equilibrium, it spreads out to form an anvil shape.

4. A faster wind aloft may move the top farther downwind from the cloud.

5. The underside of the anvil can adopt an alto or nimbo-stratus formation.

6. At times globules can form known as **mammatus (mamma)** cloud caused by the exterior skin of the cloud cooling and descending before evaporating and disappearing.

Mamma has a distinct 'udder-like' appearance which also has an ominous look about it

7. When the convection dies down and the cloud decays the tops, now formed of ice crystals, frequently remain in the sky long after the cloud itself has gone. Such cloud is often referred to as 'false cirrus' – its real name is cirrus spissatus cumulonimbus genitus! You can call it ci spi.

Ci spi is particularly noticeable when the remainder of the cloud has decayed and the top remains lingering in the sky

Cumulonimbus ascending above all the other cumulus

The cu-nimb is the most powerful of all clouds and warrants a chapter of its own later in view of the adverse effect it can have on flight safety.

The Contrail

These can be deemed to be 'man-made' clouds in the sense that they would not occur unless an aircraft was flying through the upper atmosphere at the time. A **contrail** (an abbreviation of condensation trail), is formed when aircraft engines emit exhaust containing residual moisture. This moisture immediately condenses into water particles, which in turn freeze on contact with the cold upper air. A cloud resembling cirro-cumulus forms trailing back from the source; if there are four engines then there will be four trails immediately behind the aircraft. However, they soon merge into one trail as each spreads out in the slipstream or upper wind.

The trail can be very short or seem to go on for ever. The length depends on the relative humidity of the environment at the time: if it is low the contrail will quickly evaporate into the drier air. If the relative humidity is high then evaporation may not be possible; the contrail will therefore maintain considerable length and may linger in the sky for quite some time.

A short-lived contrail soon to fade away

A lingering contrail turning into 'man-made' cirro-cumulus

These contrails are not depicting separate airways – they are highlighting a strong upper air crosswind

A momentary contrail can sometimes be seen low down coming from the wing-tips of an aircraft at an air display when the air is very damp. Such short trails are caused by a reduction of pressure at those tips, caused by the lift-producing properties of a wing and particularly accentuated in a tight turn. Just as increasing pressure causes heat – recall a bicycle pump after use – so does a decrease in pressure bring about cooling, hence the condensation that takes place during such manoeuvres.

The Distrail

Really the *distrail* is the reverse of a contrail. A high-flying aircraft may be passing through a thin veil of cirrus when its warmth causes the ice crystals making up the cloud to disappear (or, in chemical terms, sublime into vapour). The result is a clear passage behind the aircraft.

An aircraft has caused sublimation in a layer of cirrus cloud

CHAPTER THIRTEEN

PRECIPITATION

Precipitation is the all-embracing word for drizzle, rain, sleet, snow or hail. It is classified as being orographic, frontal, convection or cyclonic (convergence) according to the cause. The formation and structure of the various forms such as snow crystals can be complex; we will look at them simply as we see them.

TYPES OF PRECIPITATION

Rain

Composed of water particles which wander round and bump into each other in cloud, eventually joining up to form larger droplets or drops. When the drops become too heavy for the vertical motion in the cloud to sustain them they fall from the clouds to the surface.

Rain from strato-cumulus

A heavy rain shower from an approaching cu-nimb cloud

Rain can be defined as light, heavy, patchy or continuous according to its observed state. Under the heading of rain must come the **shower**.

Showers are associated with convection cloud (cumulus) and distinguished from rain by the relatively short length of time they are active. If one should last longer than one hour it becomes reported as continuous rain.

Another means of distinguishing a shower from rain is the intensity and size of the drops. The vigorous up-currents in large cumulus or cu-nimb can sustain the droplets for quite a considerable period during which time they can grow to a much larger size than normally found with rain.

Drizzle

Very tiny water particles which are so fine they virtually float to the ground rather than fall. Drizzle is easily distinguished from rain; it makes no splash in a puddle.

Snow

Given that conditions are cold enough, sublimation into ice crystals takes place at high level and, as they fall through, say, altostratus, any supercooled water droplets present will freeze on contact with the crystals which then become very large in a short space of time. The same action takes place in a cu-nimb cloud.

You may recall that sublimation into ice crystals cannot take place without the

A heavy snow shower approaching. Compared with the rain shower the snow shower and snow cloud in general seem to have a 'heavier' appearance in presentation and colour

presence of freezing nuclei. To do so without the nuclei, or striking a frozen surface, would require temperatures of –40°C (which is also –40°F) and below. With good freezing nuclei, ice will form at –10° to –15°C (14° to 5°F).

This 'pearl of wisdom' about freezing nuclei should at least explain why you sometimes see rain instead of snow even when the temperature around you is below freezing point at 0°C (32°F).

Sleet
As snowflakes fall, should they pass through warmer air on their descent, some of them can melt and become raindrops. **Sleet** is the resulting mixture of the snow and rain.

Hail
A dramatic but usually short-lived phenomenon. Starting as a water droplet in cunimb cloud it is carried up to great heights in the vigorous up-currents and freezes during the ascent.

On descending it collects more water droplets around it in the lower levels before once again rocketing up – this time to freeze into an even bigger particle of ice known as **hail**.

This sequence continues until the vigorous up-currents begin to weaken as the cloud reaches maturity and begins to decay, or the weight of the hailstones become too much. In either case, or due to a combination of both, they then fall to the ground. It has been known for hailstones to have grown to golf ball size and even larger, with considerable damage a possible outcome – particularly to an aircraft.

Virga

Sometimes precipitation can fall from a cloud and evaporate before reaching the surface. It can appear similar to a curtain draped below the cloud and when swept back it is known as **trailing virga**.

Virga can be seen emanating from clouds at all levels and, apart from cirrus where it would be in the form of ice crystals, it can also be composed of ice crystals from alto-cumulus – for example floccus, when conditions are right

Dew

Precipitation need not come from cloud alone. Given a clear night with no cloud cover to act as an insulator, the Earth's heat rapidly escapes into space leaving the immediate surface very cold. Moist air immediately in contact with this surface can cool and condense leaving droplets called **dew**.

Frost

In the event of freezing conditions, instead of dew forming, the change will be directly into soft ice crystals known as **hoar frost**.

Frosty field

The main flight problem with most forms of precipitation is poor visibility, and with hail, potential damage to the aircraft.

CLOUDS ASSOCIATED WITH PRECIPITATION

We next examine the cloud types associated with the various types of precipitation and will take the stratiform category first.

Condensation in stratiform cloud is a slow process usually taking place over hours rather than the matter of minutes that can be the case with cumuliform cloud.

Also, the turbulence in stratiform cloud is gentle rather than vigorous and thus the up-draughts are weaker so the droplets do not grow to any extent.

There is one exception. Nimbo-stratus in a frontal system can reach up to great height and therefore it can contain plenty of droplets ready to coalesce (join together) and become large.

Now we can look at each type of stratiform cloud that can produce precipitation:

STRATIFORM

Stratus

The water droplets contained in stratus will be very fine and to be present in any great number the cloud will need to be fairly dense. Its contribution will be drizzle.

Strato-cumulus

Dense strato-cumulus can produce periods of continuous or patchy periods of rain (or snow) ranging from light to moderate; normally this occurs only when associated with a frontal system. Drizzle can also come from strato-cu when it is less dense.

Alto-stratus

Thickening alto-stratus associated with an approaching front will produce precipitation in the form of rain or snow, which sometimes evaporates before reaching the surface.

Nimbo-stratus

Very much associated with a frontal system, nimbo-stratus is an extension of thickening alto-stratus reaching down to low cloud level. It can produce continuous moderate to heavy rain or snow over many hours.

CUMULIFORM

Large cumulus

Large cumulus, with the 'cauliflower' heads, is quite likely to produce showers of rain or snow.

Cumulonimbus

This cloud will produce just about every type of precipitation imaginable. Basically it will be in shower form if emanating from a solitary relatively small cu-nimb, but when very large cu-nimbs merge together or an individual cloud is slow moving or stationary, the outcome will be more like continuous rain – or snow – with outbursts of hail being quite on the cards.

There is a marked difference in the height to which a cu-nimb will climb in winter compared with summer – recall the seasonal variation in the height of the tropopause. Naturally this has a direct bearing on the amount of water droplets it can carry.

During the summer season it can reach up to and even penetrate the tropopause and encroach into the stratosphere. In tropical regions it can even climb to around 50,000 ft – hence the deluges experienced in monsoon seasons.

Cu-nimb reaching over 45,000 feet over New Guinea

CHAPTER FOURTEEN

ICING

We have already discussed how ice can form; now we go on to the types of icing that can relate to flight safety.

HOAR FROST
Hoar frost consists of ice crystals formed by sublimation on to a cold surface with a temperature below 0°C (32°F). It can occur on an aircraft when parked overnight or when it is descending from a level below freezing point into warmer, moister air. Such frost, or any similar deposit, must be totally removed from an aircraft prior to flight. Not only does it add extra weight but it can affect lift by distorting the smooth flow of air over the wing.

RIME ICE
Rime ice forms when small super-cooled particles of moisture freeze on striking a surface. In flight it can occur when passing through cloud which has a relatively low moisture content but containing those small super-cooled water particles. It will appear as a white opaque deposit confined to the leading edge of the wing and its immediate surround.

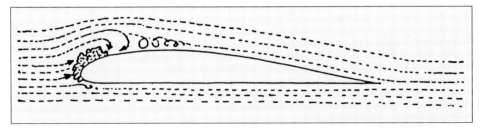

Rime ice building on the leading edge of a wing

Rime ice on the leading edge of a sailplane wing

As it builds up, some will break away in the airflow but sufficient can remain to disturb the smooth passage of air over the wings, thus affecting lift and thereby the stall speed. Rime ice is usually associated with stratiform cloud where the super-cooled particles are small.

CLEAR ICE

Also called **glazed ice**, this is the most hazardous form of all. An aircraft flying through dense towering cu and cunimb cloud will be striking very large super-cooled droplets indeed.

Initially, on striking the aircraft, only *part* of the droplet will freeze

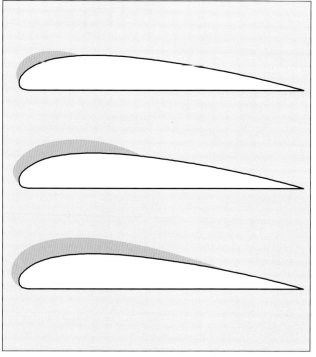

The development of clear ice on a wing

instantly. Recall how latent heat is given off when condensation takes place; so, further latent heat is released when a liquid changes into a solid. This heat momentarily delays complete freezing of the droplet. This enables the part of the droplet which is still in liquid form to flow back over a wing or any other flying surface before it solidifies.

Although it will virtually freeze as it flows it will now be covering a much larger area. With the continuous repetition of the process, as more droplets strike the aircraft, the aerofoil shape of the wing can be severely distorted – not to mention the additional weight produced. **Ice accretion**, as it is called, can have a marked effect both on the aircraft's handling and the stall speed.

RAIN ICE/FREEZING RAIN

Similar to clear or glazed ice but on this occasion the aircraft will be flying in sub-zero temperature *outside* cloud but through rain falling from relatively warmer moist air above – usually ahead of a warm front.

As already discussed under freezing nuclei in the chapter on 'Humidity', should the rain drops become super-cooled in falling through the sub-zero temperature from a warmer source above, they will freeze on striking the sub-zero surface of the aircraft.

If the accretion from rain ice looks like being a problem it may be worth climbing up into the warm air from which the rain originated to rid yourself of the ice, and then consider turning back the way you came to try again another time!

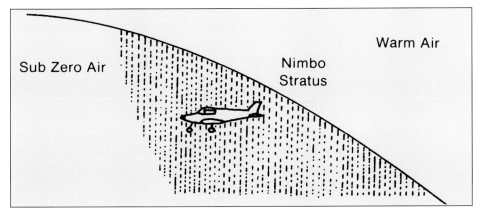

Rain falling from warm frontal air above can freeze on striking the cold surface of a wing flying in freezing conditions below

CARBURETTOR ICING

The suction which draws fuel into the carburettor brings with it a decrease in pressure due to the venturi effect (refer to lift in your notes on principles of flight), and the decrease in pressure brings with it a decrease in temperature. Add to this the further cooling due to the evaporation of the fuel and the combination can produce a dramatic reduction in temperature within a matter of a second.

Due to evaporation cooling this drop can be as much as 34°C (61°F). When the temperature in the carburettor inlet is below freezing point any water particles in the air inducted into the carburettor will initially adhere to the sides (being liquid), and immediately change into ice.

Carb icing, as it is usually called, can build up to such an extent that not only is power reduced but a complete engine failure can occur. Conditions conducive to this form of icing are very frequent. You may think that it is associated with freezing outside temperatures – it is not.

Given reasonably moist air and an outside temperature between –7° and 21°C (19° and 70°F), carb icing can occur at any time. It is the mechanics of the carburettor operation that provides the cooling – not the outside air.

The most likely situation will be when an engine is throttled back during descent. The butterfly has now narrowed the passage of air and speeded it up with the result being a further decrease in pressure

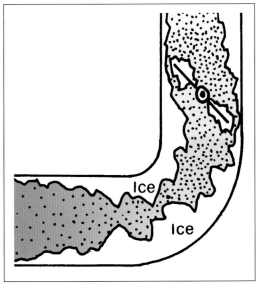

The faster airflow through the carburettor reduces pressure and therefore temperature as a result; in this case it can do so to a very great extent

Carburettor icing

(venturi effect) producing a corresponding further decrease in temperature. This calls for intermittent bursts of power during the descent.

Carb icing will not be a problem on very dry days or when the outside air is sufficiently below freezing point to produce ice crystals which, being solid, are ingested straight into the engine.

EFFECT ON FLIGHT INSTRUMENTS

Some of your basic flight instruments, the altimeter, airspeed indicator and vertical speed indicator, rely on a static pressure input to operate correctly.

With the small entry aperture to the static pressure line, icing can be a real problem when super-cooled liquid drops block the aperture and freeze.

With changes in static pressure now cut off the reading from an altimeter would also 'freeze' remaining as it was immediately before the blockage occurred.

The air speed indicator also requires a static pressure input but it needs a dynamic pressure input as well. There is an added problem here as the inlet for the latter in the form of the pitot tube has a much larger aperture.

As it faces directly into the airflow it can ingest many more super-cooled water droplets or ice particles than the static inlet. When icing conditions affect this instrument the reading could suddenly register zero.

You may consider that these problems have already been taken care of with a heated pitot head to resolve the dynamic pressure problem and an alternative static inlet placed in a heated cockpit or cabin to solve the static pressure problem. However, you may not have these niceties on your aircraft, or the electrics may fail, so you must recognise the effect icing can have on your instrument readings.

CHAPTER FIFTEEN

THE CUMULONIMBUS

Before specifically delving into the cu-nimb itself it must be remembered that *any* form of convective (cu) cloud can produce a turbulent ride for the pilot both below and within the cloud.

Any form of convection cloud can create turbulence

The extent depends largely on its size but when that anvil is forming, or has formed, you should be in no doubt as to the prospects. The cu-nimb is present and its power can at times resemble in appearance another form of awesome power.

Some of the diagrams that will follow may appear contradictory in that they show various directions for the air currents within the cloud. Sadly they can all occur at one time or another according to the cu-nimb's status at that moment – ranging from growth through maturity to decay. Never try to assess the status for yourself in order to take chances – just keep out of the way at all times.

We will now deal with the problems it can present:

Nature's atomic bomb

HAIL

As mentioned earlier, it is surprising the amount of damage that can be done by hail from a cu-nimb – particularly to an aircraft travelling at high speed.

LOCAL WINDS

In a developing cu-nimb, convergence towards the cloud takes place to replace the rapidly ascending air within it. This is how the horizontal wind shear referred to earlier is caused – resulting in different surface wind directions in proximity of the cloud.

The effect can reach out 10 to 15 miles from the cloud's location. You could note a windsock's direction and select a take-off path only to find as you accelerate down the runway that the airspeed is nowhere near reaching lift-off speed.

With cu-nimbs in the vicinity, again check the windsock at the hold point immediately prior to taking off

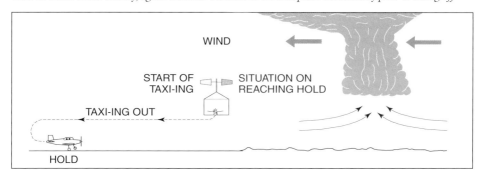

During the few minutes you were taxying out, the wind had changed completely in proximity to a cu-nimb and you were taking off downwind. Yes, this has sometimes happened with unfortunate consequences when the field has been too short.

GUST FRONT

The downdraught from a cu-nimb, descending from freezing heights and being very cold indeed, can lift up the warm surface air ahead of the cloud's path in the same manner as a cold front which will be discussed later under 'Fronts'.

Not only will you notice a temperature drop but the intensity of the wind is quite marked. This gust front is often called a line squall and very light aircraft on the ground can be turned over if not tethered down.

The gust front can be accompanied by a roll cloud formed when the cold air lifts up warmer air in its path. Should there be instability in the air at that point, further cumulus cells can be produced to enhance the cu-nimb's growth and size.

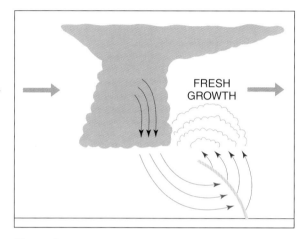

FRESH GROWTH

The gust front

A line squall in progress

MICROBURST

The **microburst** is, in a way, an extension of the gust front but is more severe. The ice-cold rain and/or hail descending rapidly in a cu-nimb cloud can drag down with it a torrent of cold air. This flow strikes the ground at considerable speed and spreads out in all directions. It brings with it a sudden and dramatic combination of vertical and horizontal wind shear – another reason why you must keep clear of cu-nimb cloud.

Should you inadvertently (hopefully never) become involved even in only a small microburst, as you enter it, say on climb out, you will experience an increase in airspeed in view of the sudden increased headwind; this could cause you to react by reducing power to compensate. But, recall vertical wind shear discussed earlier on – this is only temporary so do **not** reduce airspeed.

As you reach the centre of the microburst the increased headwind will immediately die away only to be replaced, as you enter the other side of the phenomenon, with a strong tailwind component where there will be a sudden drop in airspeed with a subsequent decrease in lift. It is therefore wise not to reduce the power on meeting the headwind component so that you are ready for the drop in airspeed that accompanies the tailwind component.

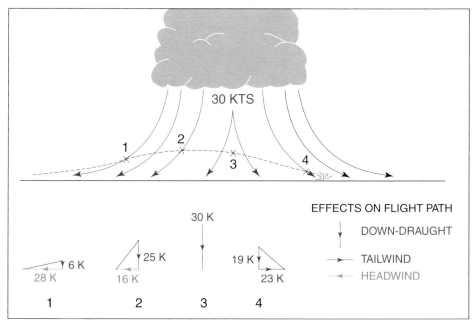

The dangers of a microburst

Finally, throughout the whole nightmare you have the major problem of the downdraught itself. This could be say 30 knots at the centre and its component can, during climb-out and approach, be enough to force the aircraft into the ground.

These substantial changes in rate of descent and the introduction of a strong tailwind component *all take place in a very short space of time* so you will begin to grasp the potential danger they present. Think about how a 30 kt microburst could affect you in terms of the varying headwind, tailwind and rate of descent components.

If the idea of altering course to escape this situation enters your mind – forget it! Apart from being unwise to turn close to stalling speed (see principles of flight) you would be doing so at just the wrong time.

When at the centre of a microburst, any escape(?) route selected would contain that substantial tailwind component which your power may be insufficient to overcome.

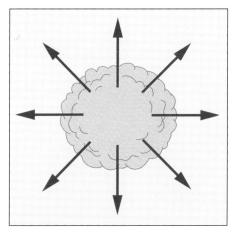

Local wind direction under a microburst

*Strictly speaking in meteorological terms a microburst of any severity is known as a **downburst** and the term microburst used only to refer to one of lesser intensity where the surrounding effect is only around 5 to 10 km (3 to 6 m). In keeping with the aviation interpretation, however, we shall stick with microburst – shades of deposition/sublimation!*

The most severe microbursts seem to occur in the Mid-West states in America, where downdraughts of 45 kt have caused the demise of airliners in sudden tailwind components of 40 kt. The very substantial power available to the airliner, not to mention the skill of the pilot, can in most cases extricate the aircraft from the situation. However, it is not easy; airliners are not able to respond rapidly to such sudden drops in airspeed because of the delay before the increased power output becomes effective – a factor common to jet aeroplanes. An added factor is that the effect of the microburst is at its maximum at low level, where take-off and climb-out/approach and landing occur.

In recent years microburst warning systems have been set up at major airfields. The problem is that such systems are on the surface and unable to give readings specific to a given height – most important in the approach stages and climb-out levels.

PERIPHERAL DESCENDING AIR

You may once again find an unexpected rate of descent has suddenly set in (recall wave effect) despite airspeed and engine revs being normal for level flight.

What goes up must come down and, although not necessarily with the same intensity, the air for many miles around a cu-nimb can be slowly subsiding. As it will be warming adiabatically there may be no condensation and you can be flying through clear air.

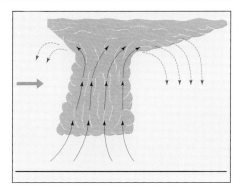

Descending air around a cu-nimb

Cold air can descend under an anvil and even further away in proximity to a cu-nimb

Just one point to remember. In the case of the first unexpected rate of descent due to wave effect, the evasive action was to move either side of your track. With the cu-nimb there is only one way to move – *farther away* from the cloud's location.

THE 'SUCKER'S HOLE'

On approaching a line of cu-nimb cloud you spot a gap and hasten to get through it to avoid a lengthy diversion of track. This is a very dangerous move – the clouds can build up very much faster than you may imagine.

You can enter that gap and suddenly find it closing round you where you and your aircraft will be subjected to all the turbulence it can throw at you.

The 'Sucker's Hole'

The 'Sucker's Hole' can close around you very suddenly — this one did in 3 minutes

ICING

The cu-nimb is notorious for icing problems, being packed full of super-cooled water droplets waiting to freeze and form clear ice on your aircraft at the drop of a hat.

Rain ice is also a possibility. Recall how precipitation can fall at upper level from alto-stratus in an approaching front, but evaporate before reaching the surface. The anvil of a cu-nimb develops into alto-stratus, and although any fall in this case is likely to be of snow, you could well be flying through as it melts just before evaporation; your ice cold aircraft does the rest.

An anvil extending for many miles from its source cu-nimb

In case you may be thinking 'I am unlikely to be flying anywhere near a cu-nimb so it is unlikely I would be under an anvil' – think again. The anvil can spread out to the extent that the source cu-nimb can be out of sight. I have seen the anvil of a cu-nimb over Southampton extend north to Oxford – around 70 miles.

TURBULENCE

Suffice it to say that the turbulence within a cu-nimb can be so horrendous that you should not dream of entering this type of cloud. The rapid up-currents immediately alongside equally rapid downdraughts can bring about structural failure.

The extreme turbulence associated with a cu-nimb is no doubt the worst of all, but turbulence in less ominous conditions can nevertheless have a serious effect on flight. To a lesser degree, a similar effect will exist in medium to large cumulus, and even flying below small cumulus can give you a roughish ride when vigorous convection currents are about.

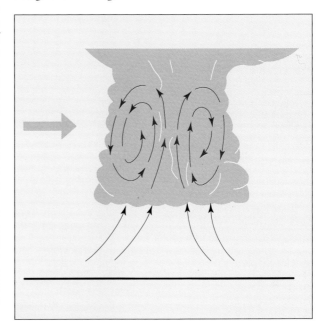

Elsewhere in your syllabus you will be taught about manoeuvring speed when you need to slow down to lessen the effect of severe vertical and horizontal gusts on your airframe.

THE 'NO-GO' CLOUD

From what you have just read you should now realize that with cu-nimb you simply do not take risks; avoid such cloud at all costs, and even pay due regard to a very large 'cauliflower' cumulus. It is more than likely to be a cu-nimb in the making.

It is, of course, the cu-nimb that is very much associated with hurricanes and tornados prevalent in certain parts of the USA. However, such conditions are not really, or should not be, associated with recreational flying. For those interested, another book by your author, *The World of Weather* covers such phenomena.

CHAPTER SIXTEEN

THE AIRMASS

An airmass is the name given to a mass of air which has basically identical properties throughout. These properties are acquired when the mass is stagnant over a given geographical location for a considerable period of time – sufficient for it to adopt the characteristics of its environment.

An airmass stagnant
- over an ocean – will become moist due to evaporation.
- over a large continent – will become dry except for relatively small moisture inputs where evaporation from lakes and rivers occurs.
- over a given latitude – will become warm at low, or cold at high latitudes.

TYPES OF AIRMASS

There are four main types of airmass to consider:

Cold and Wet	–	**Polar Maritime**	**(Pm)**
Cold and Dry	–	**Polar Continental**	**(Pc)**
Warm and Wet	–	**Tropical Maritime**	**(Tm)**
Warm and Dry	–	**Tropical Continental**	**(Tc)**

There is a fifth type which is very cold and wet. This is known as the **Arctic Maritime** airmass and it exerts its influence at times in the Northern Hemisphere.

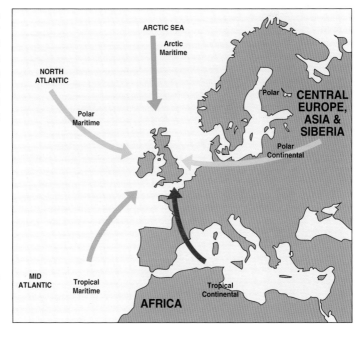

The British Isles can be affected by all five types of airmass, which tends to place them in a 'pig in the middle' situation and accounts for the ever-changing variety of weather conditions

The main source points for these airmasses are as follows:

Pm	– North Atlantic – Northern Pacific – Arctic Sea – Antarctic Sea	Tm	– Mid Atlantic – Mid Pacific – Indian Ocean – South Pacific
Pc	– Central Asia/Siberia – N.Canada/Greenland – Arctic – Antarctica	Tc	– North Africa – West Central USA (East of Rockies) – Northern India – Australia

AIRSTREAMS

When pressure systems initiate the movement of air from an airmass this air flow becomes known as an **airstream** and is named according to the nature of the airmass at its source point. Perhaps I should mention that there are times when you can come across the term 'airmass' being used to describe an airstream – a case of take your choice.

Here you can see how the general movement of air resulting from the approximate world pressure distribution identified earlier.

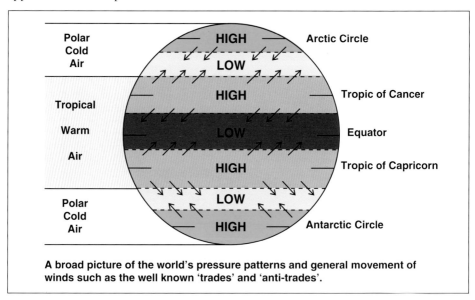

A broad picture of the world's pressure patterns and general movement of winds such as the well known 'trades' and 'anti-trades'.

The general flow from high to low pressure areas and the Coriolis Effect can be seen here

The slanting direction of the arrows are due to Coriolis Effect.

The pattern shown is a very broad picture and, as in many aspects of meteorology, reality rarely conforms to generalised patterns. The seasons of the year alone alter the picture substantially.

Airstreams have a marked effect on the weather – for good or for evil. The British

Isles can be affected by every type of airstream in existence due to its veritable 'pig in the middle' location.

This situation once led a humorist to remark that 'Britain has a delightful climate – it's the weather that's so b..... awful!'

There is an aspect about airstreams worth remembering. A tropical maritime airstream (airmass) moving north up into temperate latitudes will cool in the lower layers and produce a stable atmosphere. This associates it with stratiform (layer) type clouds.

The polar maritime airstream (airmass) heading south will warm at the lower levels and can create instability which in turn may trigger off convection. In this case the association is with cumulus (heaped) cloud.

TRANSFORMATION

In passing from one location to another, given that enough time is to hand, an airstream can take on the characteristics of the areas over which it travels.

Consider the potentially unstable polar maritime airstream. Should its path extend too far south in swinging round a depression, it will not only warm in the lower layers but, given time, it can do so to a greater height.

In fact to a certain extent its state can change into that of a semi-tropical maritime airstream. When this happens the airstream is called a **returning polar maritime airstream**. The weather can be very different with at least the lower layers changing from potential instability to stability.

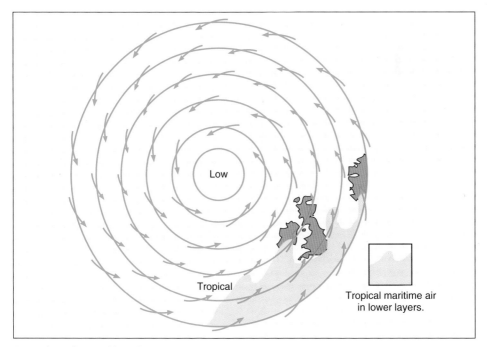

Returning polar maritime airstream

This process of change in an airstream is called **transformation**.

CHAPTER SEVENTEEN

DEPRESSIONS

A depression, colloquially referred to as a **low**, exists when there is a reduction of air pressure at the Earth's surface over a given region. It will possess a defined centre, as depicted earlier when pressure systems in general were identified.

Recall our touching upon convergence when we discussed the way in which air could ascend to form cloud. Air in the upper levels of the troposphere can be caused to speed up and leave a region – its movement is a process known as **divergence.** The air following on behind can be taken by surprise and be slow to catch up. So, in the interim, air rises up from the surface to fill the 'gap'.

The ascending air reduces the pressure at the surface and a gradient force is initiated which sets up a flow of surface air into the region to replace it; this process is known as **convergence**.

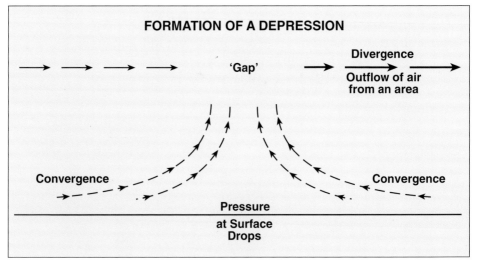

Convergence at the surface due to divergence in the upper air

Surface pressure will continue to lessen as long as the divergence aloft remains greater than the convergence thus sustaining a depression at the surface.

Precisely where and when a depression is going to develop is still a mystery – an example of why the foibles in meteorology make it an inexact science.

Weather associated with a depression is invariably depressing(!); it depends on its depth and the closeness of the isobars around the system. As a norm, a deep depression will bring strong winds, and even gales of considerable severity in exposed places.

Rain will be virtually continuous from nimbo-stratus which, with associated fracto-nimbus, can hang around at very low level for quite some time – all impeding visibility. Even thunderstorms can occur from cu-nimb embedded in the nimbo-stratus should the air becomes unstable.

We will now look into some of the types of depression that exist.

POLAR DEPRESSION

The Polar Depression forms along the line where tropical maritime air (warm/wet) comes up against polar maritime air (cold/wet) in the temperate latitudes. The line of this meeting point is known as a **quasi-stationary front**.

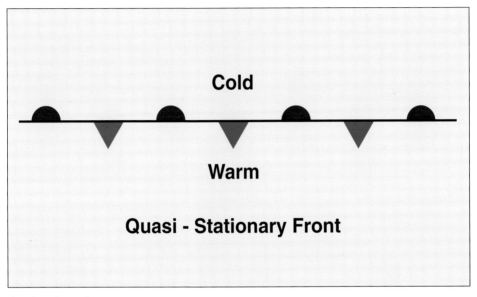

Quasi-stationary front

The finest examples of the polar depression are to be found in the North Atlantic. Born off the Eastern seaboard of the North American continent they travel across the ocean penetrating well into the European continent before dying out.

The divergence to set up the necessary surface convergence for the depression to form comes from faster flowing upper winds; at the level concerned there can be jet streams. Incidentally, their direction can also be different from surface winds over the same location.

THE THERMAL DEPRESSION

There is another way in which a low pressure area can come into being. It is rather similar in origin to the initial cause of sea breeze effect, but this time it occurs inland and even over a whole continent.

With a very hot summer day a region can heat up to the extent that a large mass of air can begin to rise, and it will be more than just a simple thermal caused by local uneven surface heating.

The ascending air will reduce the surface pressure and a mini depression can result over the region known as a **thermal depression**.

Similar to the sea breeze, this mini depression will only last as long as the sun is up to provide the necessary heat. When it comes to the case of occurring over a whole continent, however then it is not just a daily event but one which can extend over months.

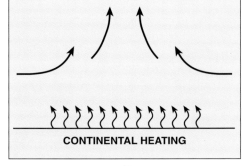

An upper air pressure gradient

Just from an interest point of view, there is a case where such a depression does not fade away with the setting sun. In the tropics the land can heat up so much that, despite radiation at night, it can be warmer at the beginning of the next day than it was at the beginning of the previous day, thus leading to the progressive warming of a whole continent over a relatively long period.

An example of this is the Indian sub-continent when the air eventually rises in general over the whole country and results in the convergence of moist air from the Indian Ocean. This is the cause of the Indian Monsoon and the continuous heavy rains associated with it that can last for months.

THE OROGRAPHIC DEPRESSION

This type of depression can form when air piles up against the windward side of a mountain range which then acts as a form of dam. The reduction of airflow to the leeward side can in turn cause a reduction in pressure and the formation of a local depression.

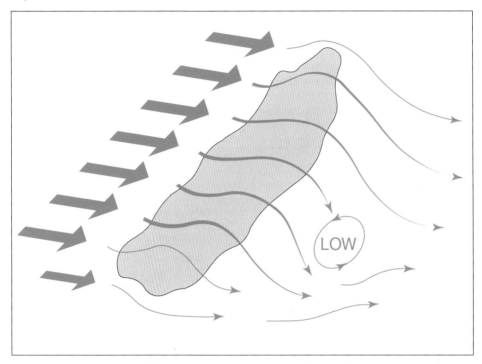

The 'dam' effect of mountains builds up pressure on the windward side thus reducing it on the lee side

CHAPTER EIGHTEEN

FRONTS

Frontal systems in the temperate latitudes are normally associated with polar depressions and form along the line where tropical maritime air (warm/wet) comes up against polar maritime air (cold/wet). These airmasses appear like two opposing armies at the front in World War 1. This analogy led to the adoption of the term **front** to describe their meeting place.

As long as no activity takes place between the warm and cold air, the line between them is the quasi-stationary front referred to earlier.

When a polar depression forms along this line the circulatory motion that sets in creates a 'kink', or wave, which sees the warm air encroach into the cold air. This process continues and produces a distinct **warm front** and **cold front** which are shown on a chart by red semi-circles for the warm front and blue triangles for the cold – both placed ahead of the airmass they each represent.

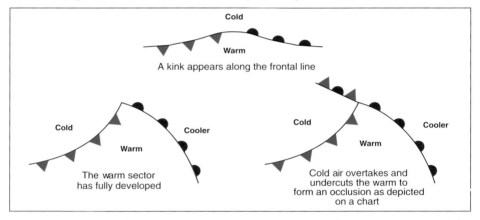

As the system develops two distinct sectors are formed; the **warm sector** and the **cold sector**.

From a surface cross-section viewpoint, notice how the warm (less dense) air ascends over the cooler (denser) air ahead. In turn, the faster colder air following behind the warm air undercuts it and forces it to ascend. The slope of the warm front is about 1:150 whereas that of the cold front is only 1:50.

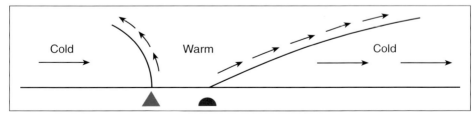

A cross-section of a polar frontal system

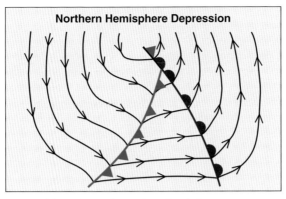

A frontal depression in the Northern Hemisphere

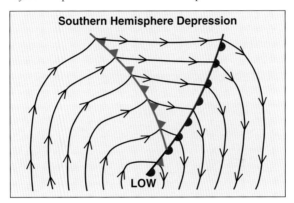

A frontal depression in the Southern Hemisphere

As a rough guide, a warm front travels at about two-thirds of the geostrophic (2000 ft plus) wind whereas a cold front is faster at the speed of the geostrophic wind itself. The relationship of the fronts to a depression appears thus in the two hemispheres.

Note how the isobars 'kink' as they cross each front; this shows that as a front passes over a location the wind veers in direction in the Northern Hemisphere and backs in the Southern.

The cloud pattern shown diagrammatically is that pertaining to what is known as a 'classical' frontal system.

However, rather like the standard atmosphere the classical front only represents an idealised pattern; frontal systems in reality can vary and thus the associated clouds and weather conditions can do likewise.

Earlier, when discussing wind, reference was made to the fact that there could be a vertical motion in the atmosphere as well as a horizontal; albeit this was slight unless attributed to vigorous convection.

Suppose the overall atmosphere around a frontal system were rising slightly. It would be cooling adiabatically along with the air being forced to rise at the front. The cooler it becomes the more likely it is that condensation will continue to take place and in this situation the clouds can reach up to considerable heights.

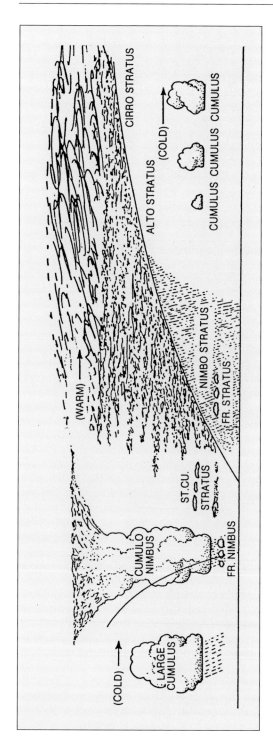

The 'classical' frontal system

Fronts associated with rising air are known as **anabatic (ana) fronts**. The associated rain extends to around 240 miles (400 km) ahead of the warm front and recurs at around 60 to 120 miles (100 to 200 km) ahead and/or behind a cold front.

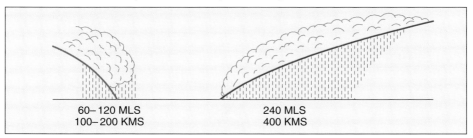

Precipitation pattern in a frontal system

We now come to another type of frontal system. When the overall atmosphere is descending it will be warming at the adiabatic lapse rate, and the warm air being forced to rise at the front will more quickly reach equilibrium. This prevents any further climbing and further cooling, so the height to which condensation and therefore cloud can reach will be reduced.

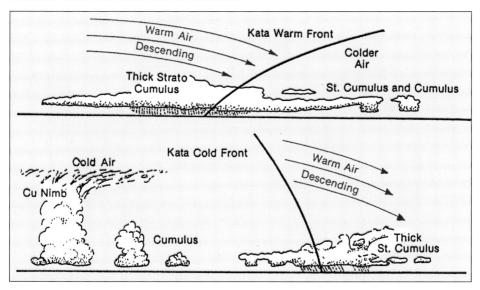

A kata frontal system

Fronts associated with descending air are known as **katabatic (kata) fronts.**

The pattern is less well defined compared with an ana system. However, both do follow the broad principle of thickening cloud from a westerly quarter culminating in a clearance from the west as the system passes through.

The weather pattern associated with a frontal system would be broadly as follows.

A picture speaks a thousand words so we will now have a look at sequences of both ana and kata types of front to see what they can really look like in actual pictorial form.

ANA FRONTAL SYSTEM

The Approach

Weather pattern in a passing frontal system as seen by an observer

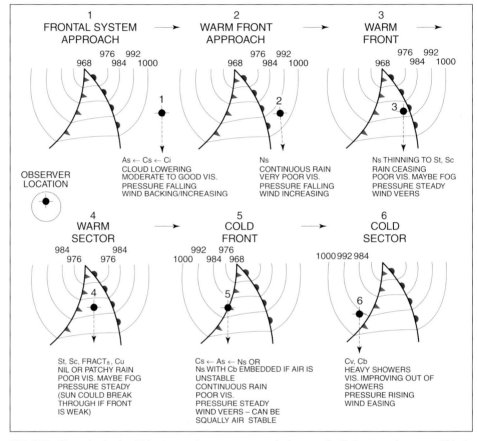

1 FRONTAL SYSTEM APPROACH →
976 992
968 984 1000

OBSERVER LOCATION

As ← Cs ← Ci
CLOUD LOWERING
MODERATE TO GOOD VIS.
PRESSURE FALLING
WIND BACKING/INCREASING

2 WARM FRONT APPROACH →
976 992
968 984 1000

Ns
CONTINUOUS RAIN
VERY POOR VIS.
PRESSURE FALLING
WIND INCREASING

3 WARM FRONT →
976 992
968 984 1000

Ns THINNING TO St, Sc
RAIN CEASING
POOR VIS. MAYBE FOG
PRESSURE STEADY
WIND VEERS

4 WARM SECTOR →
984 984
976 976

St, Sc, FRACT₈, Cu
NIL OR PATCHY RAIN
POOR VIS. MAYBE FOG
PRESSURE STEADY
(SUN COULD BREAK
THROUGH IF FRONT
IS WEAK)

5 COLD FRONT →
992 976
1000 984 968

Cs ← As ← Ns OR
Ns WITH Cb EMBEDDED IF AIR IS
UNSTABLE
CONTINUOUS RAIN
POOR VIS.
PRESSURE STEADY
WIND VEERS – CAN BE
SQUALLY AIR STABLE

6 COLD SECTOR
1000 992 984

Cv, Cb
HEAVY SHOWERS
VIS. IMPROVING OUT OF
SHOWERS
PRESSURE RISING
WIND EASING

BELOW: *Cirrus in the sky thickening to cirro-stratus towards the west. Declining cumulus may still be in evidence overhead*

ABOVE: *Cirro-stratus forming overhead. Cumulus has died out as the onset of the front dampens down convection*
BELOW: *Alto-stratus overhead with the sun becoming opaque or 'fuzzy' in outline*

Arrival of Front

Thicker alto-stratus to the west and lowering in height. Fragments of strato and fracto-cumulus may exist

Layers of thick alto-stratus now overhead with the sun finally obscured. Rain starts to fall and cloud base continues to drop

ABOVE: *Sky covered with shapeless nimbo-stratus of ill-defined base. Rain can be continuous and heavy*
BELOW: *Fracto-nimbus can appear below the nimbo-stratus as the air becomes saturated by the rain and lifts in turbulence to condense. The rain can continue for several hours*

Passing of warm front

Watch for a veer in the wind with the rain easing off and becoming patchy. The sky appears lighter and shape comes back into the cloud cover

Within the warm sector

Usually strato-cumulus and stratus which can vary in thickness. Light rain or drizzle at intervals. Sometimes this layer can thin out to let the sun through

Approach of cold front

The strato-cumulus and stratus begin to thicken up and the reverse of the warm front takes place

The undercut warm air once again produces nimbo-stratus with continuous rain. However, a cold front covers only approximately a third of the ground of a warm front, so its passage takes much less time

Passing of an Ana cold front

A further veer in the wind direction. The rain stops and a gradual clearance of cloud takes place to reveal blue skies with the receding front very distinctly marked by a sharp edge right across the sky from horizon to horizon

Signs of cumulus and possibly cu-nimb beginning to form in the west. Also a drop in temperature

To the east the receding ana cold front can maintain a straight line from horizon to horizon

There is a variation in an ana cold front when the air is not only ascending but also happens to be unstable. In such a case convection can be triggered off in the warm air as it is lifted by the undercutting cold air, resulting in cu-nimb cloud rather than nimbo-stratus. In fact the front can simply be a line of cu-nimb cloud producing heavy rain or hail and even thunder. The cu-nimb clouds may not be immediately apparent, as during the front's passing they can be hidden by the other clouds. However, the nature of cu-nimb-type weather usually ensures easy recognition.

The intensity of the rain, frequent attendant squalls and even thunder will soon make it clear that an unstable cold front is going through. Invariably, such a front also has a sharply defined edge at the rear, but this time you will see a line of cu-nimbs instead of cirro-stratus

The cold sector

The colder air in this sector will be the polar maritime airstream. As it travels south over a relatively warmer sea, its lower layers will become warmer. With the cold air above, this can trigger off convection, so cumulus and cu-nimb are usually the norm with showers.

The cumulus and/or cu-nimb will produce blustery showers of rain, sleet or snow according to the time of year

KATA FRONTAL SYSTEM

Approach of front

The first sign is increasing thickening strato-cumulus from the west

Complete cover of strato-cumulus

Arrival of front

Rain is now falling, being either continuous or patchy but rarely heavy

Warm sector

The strato-cumulus begins to thin out in the warm sector

Cold sector

The strato-cumulus once again thickens as the cold front approaches

Precipitation again commences from the thick strato-cumulus

ABOVE: *Once again the strato-cu thins out as the cold front passes*
BELOW: *The clearance of a kata cold-front is not so well defined as that of the ana type*

OCCLUSIONS

There comes a time in most systems when the faster cold (polar) air overtakes the point where the warm front is at the surface. It then under-cuts the less dense warm sector and raises it aloft as the less dense cooler air ahead of it is also undercut.

The event is known as a **cold occlusion** and is normally associated with the decline of a depression. It is shown on the charts as a purple line following on from the cold front line with alternating red semi-circles and blue triangles.

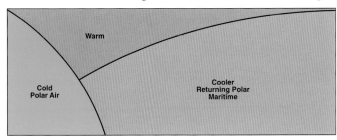

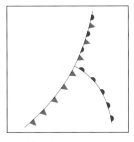

Cross section of a cold occlusion

A cold occlusion

There are occasions after a very cold spell that the cold front polar air may be *less cold* than the preceding cold air in the system. This colder air may be the remains of an Arctic airstream. The cold front air will now not only undercut the warm air but it will *ascend* over the even colder air; this event is known as a **warm occlusion**. It appears in the same colouring on a chart but the purple line will follow on from the warm front line.

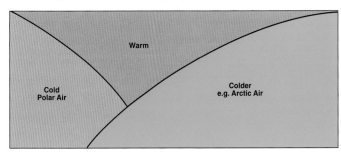

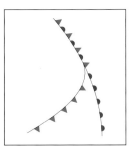

Cross section of a warm occlusion

A warm occlusion

Although the frontal system will be dying out when occlusions occur, they can still give a pretty good account of themselves. Frequently they bend back around the low pressure centre and produce a repeat performance of miserable weather.

Cloud-wise an occlusion sequence could be cirrus - cirrostratus - altostratus - nimbostratus - altostratus - cirrostratus with no warm sector experienced on the surface and no distinct veers in wind direction. There are however times when cu-nimb can be embedded in the nimbo-stratus of an occlusion.

FURTHER FRONTAL DATA

Secondary (Polar) Depression

A further polar depression can often form along the tail of a cold front emanating from the centre of the main depression. This is known as a **secondary depression** and at times a chain of such depressions can form along the front.

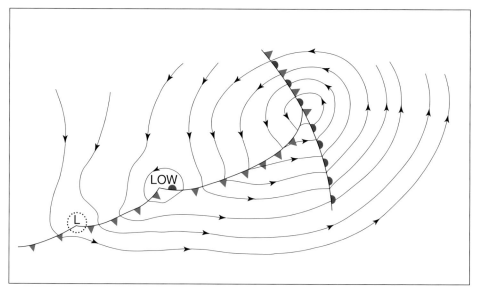

The formation of a secondary depression

'Solo' Fronts

A front need not be associated with a depression. Airstreams such as Arctic or Polar Continental air can swing round a pressure system and introduce very cold conditions bringing similar weather to that normally associated with a cold front but an even colder temperature due to their source point.

In the same manner, Tropical Continental air can move into the temperate latitudes to introduce very warm conditions. Again, the weather could be similar to the warm front we know, but there is the likelihood of it being accompanied by thundery rain, and on passing it can leave humid sticky air with resulting fog.

In both cases there will be no defined warm/cold sectors. The change would simply be a gradual one into colder air depicted by the appropriate symbol for a front on a chart. Also, they will not necessarily come from a westerly direction. The solo cold front can approach from a northern or easterly quarter; the warm front coming from a southern or south-easterly quarter.

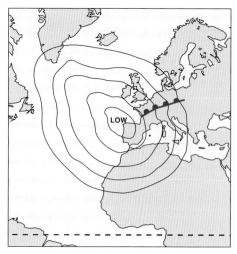

The low pressure area circulation is drawing up a tropical continental airstream of a much higher temperature than the air in front of it

Frontal thickness

Fronts need not necessarily be solid throughout. At times they can be layered with clear gaps in between. The solo front just discussed can pass through as no more than a layer of cloud if it is weakening.

A warm front is not necessarily solid – it can be composed of layers aloft

Sea breeze fronts

The cool air coming in off the sea in a sea breeze can under-cut the warm air inland, and if it is lifted into an unstable atmosphere a line of cumulus cloud will develop. This line can be seen moving slowly inland during the day, only decaying when the sea breeze itself decays at the end of the day.

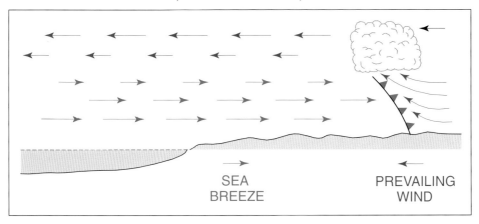

SEA
BREEZE

PREVAILING
WIND

A sea breeze front

A sea breeze front moving inland

Hoax fronts

There can be hoax fronts when the western sky is thickening but, finally, nothing happens. It is worth a barometer check to confirm that pressure is indeed falling before making a judgement. Here is an example of a front that never was!

Vigorous front

When a front is very vigorous with a steep pressure gradient, the change in wind direction as the front passes can be relatively sudden. The same can happen with pressure change from falling to rising. In fact the pressure trace on a barograph can appear almost as a 'V' instead of the normally wide 'U' shape.

There is always a snag to everything. Here is an example of a frontal system that never happened as expected. It was passing far to one side and a long way away. A pressure check would have revealed this, but a barometer was not to hand at the time

A VITAL FLIGHT TIP

If considering flying for an hour or so and a frontal system is approaching, fly *towards* it until the weather deteriorates. You at least know the route home is clear – you have just been through it.

Go the other way because it looks brighter and you may return to find a closed in airfield – not a clever situation to be in.

Upper wind direction at fronts

The upper winds around cirrus level associated with a frontal system can differ from those at the surface. The difference can be observed when visibility permits by the movement of high cloud in relation to low cloud.

General Rule:	Warm front passing	– surface	wind	south-west
		– upper	wind	north-west
	Cold front passing	– surface	wind	north-west
		– upper	wind	south-west

FINALE ON FRONTS

Without wishing to confuse or depress you, the inexact science of meteorology dictates that rarely will one frontal system be the same as another. Fronts can not only be either ana or kata – they can be a mixture of both. They can also be weak or strong and on occasions even approach from other directions than the westerly quarter. It really is a difficult world we live in!

Later we shall be looking into the met information available to pilots. The above is sufficient reason for full use to be made of this service and not to rely on personal interpretation alone.

CHAPTER NINETEEN

THUNDERSTORMS

Thunderstorms develop by the rapidly moving hail and ice particles in a cumulonimbus cloud causing positive and negative electrical charges.

A discharge to the ground is visible in the form of **fork lightning**, as it is commonly known. When the discharge is between two clouds, although the lightning may not be directly visible the whole cloud can light up from within to produce what is generally described as **sheet lightning**. The intense heat produced by lightning causes a sudden dramatic expansion of the surrounding air which becomes very audible as **thunder**.

You have already been warned about the vicious conditions in a cu-nimb so it is very necessary that you are able to recognise the approach of thundery conditions for the sake of flight safety.

Many such storms develop at the end of a summer heat-wave accompanied by poor visibility due to haze. This condition can make it difficult to see the approach of a thunderstorm until it is well developed – in fact thunder may be heard before anyone bothers to notice a change in the sky.

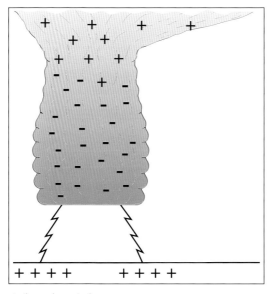

A charged cu-nimb

The haze on a summer's day can hide cu-nimb developing into a thunderstorm

There are two types of thunderstorm; low level and medium level. Here is the sequence of a low level thunderstorm; it had come and gone within an hour or so with very little warning. More than likely it was associated with a small mini cold front.

LOW LEVEL THUNDERSTORM SEQUENCE

Mid-afternoon, fair weather cu in the sky, by this time the cu should have substantially developed if the day was to produce problems.

Cu and fracto-cu

Cu and fracto-cu still around with large cumulus beginning to appear. Cirrus to the south not as yet identifiable as cirrus spissatus (ci spi) from the anvil of a cu-nimb.

Cu and fracto-cu, distant ci spi cugen in the upper air

Shortly afterwards large cumulus became more evident with a background of now obvious ci spi and a darkening sky towards the south.

Distant cu-nimb now in sight

The sky became even darker to the south with large cumulus dominating the sky and cu-nimb immediately following on.

At least one pilot was wasting no time coming home!

Cu-con appearing on the scene

That a storm was soon to break was obvious – the 'summer day' had now disappeared into oblivion.

Cu-con growing into cu-nimb

As the sky became really black the sound of distant thunder was heard, confirming that a thunderstorm was in the offing.

Distant thunder and lightning

By now the event had drawn a few spectators even though heavy rain was just commencing.

The storm breaks overhead, accompanied by vivid lightning and heavy rain

The storm broke overhead with torrential rain, vivid lightning and thunder. Sadly, the chances of capturing daytime lightning with a camera are virtually nil!

The rain continues, but eases off as the storm passes on its way

Fortunately no aircraft were airborne at the time the storm broke – but they had been only a short time before.

Not all thunderstorms are caused by low level cu-nimbs resulting from uneven surface heating, or air being lifted into an unstable layer orographically or at a front. Sometimes, in summer, a pool of cold air which can be associated with an upper air trough can create an unstable atmosphere. Storms ensuing from such instability can build up to reach the Tropopause after starting at medium level; in this case there is usually plenty of warning.

When previously discussing medium cloud, mention was made of two types of cumuliform cloud at this level which could portend thundery conditions – let's take the first type:

Alto-cumulus castellanus (Ac cas)

Alto-cu cas consists of a series or rows of small hard turreted-top cumulus with the traditional well-defined flat base but forming at medium cloud level.

ABOVE: *Castellanus consists of a series of small, hard, turret-topped cumulus with a well-defined flat base*
BELOW: *Another display of Alto-cu cas*

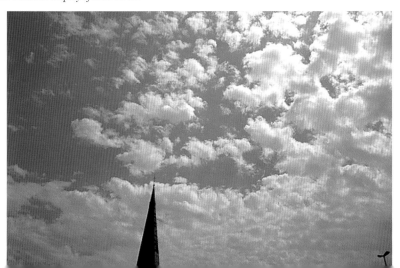

Alto-cu cas is indicative of instability at medium level and there will usually be thundery conditions developing within a radius of 100 miles if not at the point where it is observed.

Here is a picture taken in North Oxfordshire and showing where 75 miles away over Cardiff in South Wales, apart from the cas cloud, you can just see the top of a cu-nimb. Later investigation revealed that a heavy thunderstorm had occurred from this build up 75 miles away, but nothing happened at the source point of the observation.

This castellanus proved to be an omen of potential thundery conditions. In the far distance cu-nimb can just be seen producing a storm 75 miles away

The other type is:

Alto-cumulus floccus (Ac flo)

Alto-cumulus flo consists of irregular patches of cumulus cloud at medium level from which emanate streams of trailing virga. Although not at high level, this virga can be composed of ice crystals. Flo cloud can also occur at high level as cirro-cu flo.

Alto-cu flo

As just intimated, alto-cu cas or flo can make an appearance and simply disappear without any thundery conditions developing at the location where it is observed. But at least they do give reasonable warning of the possibility.

Now follows a sequence depicting a medium level thunderstorm:

MEDIUM LEVEL THUNDERSTORM SEQUENCE

First signs of alto-cu cas in the sky.

First signs of castellanus in the sky

The alto-cu cas has increased.

The castellanus increases

There is now a substantial cover of cas in the sky.

There is now a substantial coverage of castellanus

The sky is now totally covered with the alto-cu cas, hiding the cu-nimb developing from it.

The sky becomes totally covered and hides the cu-nimbs beginning to form above the cas layer

Thunder and lightning commence with heavy rain occurring 9 miles to the west. The sky in that location has developed a nimbo-stratus appearance.

Thunder and lightning commence with heavy rain only 15 km (9 miles) to the west

The storm continues on its path northwards having missed the observation location on this occasion

Thunderstorms can also be associated with occlusions when the warm air is lifted up into colder levels and convection is triggered off.

CHAPTER TWENTY

HAZE, MIST AND FOG

HAZE

Haze is composed of minute dust and/or smoke particles which are so small that they remain suspended in the air and only settle on the rare occasion of a nil or next to no wind condition when no turbulence is present.

Haze

The formation of haze is quite straightforward; there is no physical process involved. People burn fires world wide and dust particles can be lifted up by convection or convergence to considerable heights and then be carried for great distances by strong upper winds. Powdery sand from the Sahara Desert has occasionally been deposited in the UK.

MIST

Mist is no more than very thin cloud on the surface. Given a high relative humidity and cold air, condensation soon occurs.

When the dew point temperature is the same as the surface temperature, the surface air will be saturated and the slightest further cooling will produce condensation.

Mist is composed of very fine water particles suspended in the same way as haze; they are, however, larger than haze particles.

Mist

FOG

Fog is a development of mist that occurs when the water particles have become very much larger due to continuing condensation.

Fog rarely forms without going through an initial mist stage unless it is blown in from elsewhere 'ready-made'. A heavily polluted atmosphere will provide an abundance of condensation nuclei.

Internationally, fog is considered present when visibility is 1 km (1100 yards) or less

TYPES OF FOG

Radiation fog

After the sun has set, heat will be given off by radiation overnight. Add to this the absence of any insulating blanket of cloud and the ground can soon become very cold; at this stage, so will the air in immediate contact with it.

Given a high relative humidity, a light wind and just enough turbulence to lift and mix this air, fog begins to form.

Fog brought about in this way is known as **radiation fog**, and it is most prone to occur on autumn and winter evenings and on spring and summer mornings when the sky is clear – probably before anyone is up in the latter case!

Here is a sequence of early morning radiation fog forming.

Dispersal of radiation fog takes place when the sun warms the ground through insolation, and the air in turn is warmed sufficiently for the fog to evaporate within it.

During the winter season the sun is low in the sky and the day is short so insolation is at a minimum, making it slow for fog to clear.

At times it can lift into low stratus and then disperse to reveal the sun; but if the warming is insufficient, say in winter, it can remain as low stratus all day.

A sequence of radiation fog forming

For radiation to have a real effect a clear sky is necessary so that the heat can escape rapidly through the atmosphere and eventually into space. Any cloud cover can act as a blanket and prevent sufficient heat being lost for the air temperature to fall below dew point

As fog disperses it usually lifts to form low stratus cloud before breaking up completely

Be warned, radiation fog can form very rapidly given a clear sky on a late autumn or winter afternoon when water droplets appear on your cold car to indicate high relative humidity. A friend of mine suggested we go flying on such a late afternoon. In refusing the offer I explained why. The next morning my reasoning was sadly confirmed by the announcement of two separate fatal accidents during attempts to land in fog the previous evening.

It really is not worth it.

Hill fog

When you stand in **hill fog** it will appear to be no different from ordinary fog; however, viewed from the plain below it shows its real identity as hill stratus. It is particularly prevalent in coastal regions where it forms orographically as moist air coming in from the sea meets high ground inland.

Advection fog

This is simply fog which has already formed elsewhere and drifts from one location to another, as touched upon earlier; it resembles a bank of cloud on the surface as it approaches.

Coastal airfields can suffer from **advection fog** blowing in off the sea. It is caused when warm air coming up north from the tropics is cooled on passing over a cold sea in the temperate latitudes and condenses. Good examples of this are to be seen off the southern shores of Cornwall.

Hill fog is really stratus cloud covering the surface of the hill

Advection fog rolling in on a light wind

CHAPTER TWENTY-ONE

THE ANTICYCLONE

Usually referred to as a high, the **anticyclone** is best known for the weather it does *not* produce although it can have a marked effect on visibility. It is formed in the opposite way to a low (depression) with the upper air converging on a region and subsiding with divergence taking place at the surface.

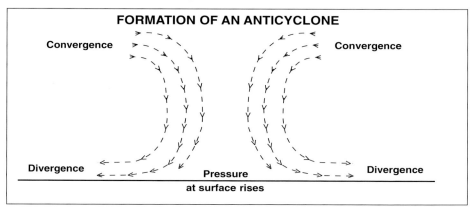

The formation of an anticyclone

During summer it is a high with its associated stability which brings those days of blue skies with unbroken sunshine and light winds when near the centre.

A typical anticyclone

Inversions are very much associated with **anticyclones** where the subsiding air warms adiabatically during its descent. The environmental lapse rate will be low so any air forced to rise will almost immediately become cooler than its environment – recall ELR < ALR meaning stability.

The inability of air to rise in an inversion means that haze, mist, fog and industrial pollutants can be trapped until the inversion breaks down. Should the inversion be caused by an anticyclone the entrapment can last for days.

A frequent sight in the early evening — smoke unable to rise through an inversion

Take the same conditions during the winter when the sun is low in the sky and the day is short; insolation is at a minimum making it difficult for the sun to clear any fog.

On the other hand, there may be no fog but a layer of low cloud – usually stratus. This may or may not break up during the day through lack of warming at this time of year. When a thick layer of such cloud refuses to budge it gives rise to a very depressing format known as **'anticyclonic gloom'**. It has the amorphous appearance of nimbo-stratus but no rain.

In climbing flight, on reaching the top of an inversion there is a distinct line, and having broken through it you will suddenly find yourself in completely clear air.

Sometimes a parcel of air which is substantially heated over and above its

Note the clear dividing line where visibility improves at height

An example of a situation in which convection has 'punched' through an inversion could be the heat generated by power-station cooling towers

surround may just keep warm enough to climb and punch through the inversion, where convection may be triggered off. Any resulting cumulus is unlikely to grow to a great height due to the overall stability of the atmosphere in such conditions.

When an anticyclone sets in over northern Europe it can bring intensely cold polar continental air from Siberia forming a bitter east wind over the UK. If you are located on the edge of the system you will recall that the winds can be quite strong around an anticyclone. But at least such winds can preclude the formation of fog and sometimes break up low stratus to give a day of clear sunshine to alleviate the chill.

The ridge between two frontal systems is transient compared with a high and there is unlikely to be any haze

RIDGE

There are two types of **ridge** – the transient one, usually a short-lived wedge or tongue of high pressure situated between two frontal systems. Of brief duration it does not normally have the time to build up the characteristics of a high such as an inversion; but it can produce cloudless skies and on this occasion there will be no haze and therefore good visibility.

Such a ridge will probably come in with a north-west wind – the legacy of a recent cold front. Its departure will be signalled by the backing of the wind to south-west, with the possibility of another frontal system indicated by cirrus and cirro-stratus to the west.

The other type can be a larger tongue extending from the high and staying around for a rather longer period of time, with its strength diverting fronts around it. It will be similar in visual terms to the anticyclone itself and will produce similar weather conditions.

COL

Not strictly speaking a pressure system, a **col** is the region situated between two highs and two lows within which winds can be light and variable.

In summer, given hot humid air, thunderstorms are a possibility; in winter the outcome is more likely to be fog.

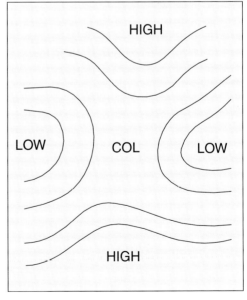

A col

CHAPTER TWENTY-TWO

VISIBILITY

Visibility is expressed in kilometres or metres with the latter usually applying when it is poor.

By international aviation standards fog is said to exist when visibility falls below 1 kilometre (3300 feet) or just over half a mile. However, in the UK as far as the general public is concerned a fog warning is only issued when the visibility is likely to fall below 180 metres (600 feet). This highlights the distinction set between flying and driving!

In the USA fog was simply listed as one of many 'obstructions to vision' – defined by the letter 'F' in reports when the visibility fell below 6 statute miles. However, with the changing to international terminology it is now reported when visibility falls below five eighths of a mile.

Flying in poor visibility is a non-starter other than for pilots possessing the experience and the appropriate rating attached to their licence. However there are other aspects of visibility which can affect the pilot who is flying simply for pleasure or recreation purposes within the privileges of the licence.

Here are some aspects possible inflight and without warning:

- With the sun behind you (down sun) on a clear day all might be well, but given, say, scattered clouds, the shadows caused on the ground could easily obliterate scheduled way-points en route.

Shadows cast by cumulus clouds can virtually obliterate en route way-points

This cloud had only to move a little to the north and London (City) Airport would have disappeared in shadow

- Heading into the sun (up-sun) will greatly reduce your visibility due to the reflection from haze or any slight mist; this is apart from the glare.

Flying up-sun can play havoc with your visibility

- The situation can be made worse when an inversion is present – usually associated with an anticyclone. Above it there will be a crystal clear sky but this won't help if you need to be in constant contact with the ground.

The following sequence was taken during a very weak inversion; it can be left to your imagination what it would be like when substantial.

Climbing up though an inversion sees a gradual reduction in visibility

A source of poor visibility near the surface is always a potential problem. You may have taken off in a visibility of 3km (2 miles) but at your destination it could be only 1.5km (4,900ft), just under a mile. This could mean that from around 5,000ft and above, your destination will not be in view.

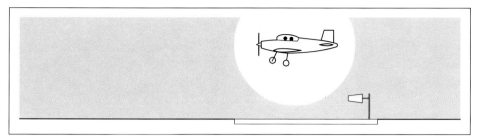

Airfield out of site overhead

- Again, you could arrive overhead your destination with the airfield just in view from 5000 ft. You descend to circuit height and on base leg into the final approach find that it has disappeared. Your horizontal distance from the airfield on approach has become greater than the 1.5 km visibility prevailing at that point.

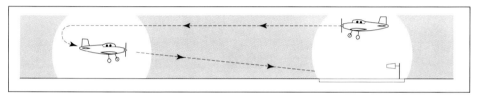

Airfield out of sight on approach

   In such conditions of poor vertical visibility you should descend (within reason) to a level where it improves.
- It goes without saying that any precipitation in the form of rain, sleet, snow or hail will cause a marked deterioration in visibility.
- Precipitation at fronts is usually continuous so very poor visibility can exist over a long period and considerable distance. Of course this will vary according to the direction in which you are heading in relation to the front's movement.
   Going with it will extend the time you experience the poor visibility; going through it should reduce that time.
   Having said that precipitation will substantially reduce visibility, it should be stressed that in all probability it would be illegal to fly in such conditions without an additional rating to your licence or certificate.
- Likewise, visibility will be very much reduced when flying at night – for which of course you need a night rating added to your licence in the UK.
   However, there is an element of assistance at night for a pilot in the form of well lit towns that may be on the route – together with moonlight when its about.

Please do not dismiss visibility problems by saying 'I've got GPS'. Batteries can run down at the most awkward moment. Perhaps when this happens, and no flight plan on a map is to hand, the look on a pilot's face will say more than any words in a book!

GPS is a brilliant invention which can greatly enhance flight safety; but there is no either/or about it for the basic pilot. The priorities are:

G	**P**	**S**
Eyeball plus Map First – Global	Positioning	*Second* !!

CHAPTER TWENTY-THREE

REPORTS/FORECASTS

Poor weather is frequently associated with low pressure – particularly in the UK where one can be subject to a string of depressions. Consider investing in an inexpensive barometer or if you can stretch to it – a barograph. Knowledge of the pressure trend at the time can supplement the assessment you make of the sky and save you a wasted journey to your airfield.

But, you cannot accurately assess certain weather situations without help. You may intend to fly to a destination many miles from your home base so there is the need to know what the weather will be *en-route* and at your intended destination. Pay due attention to the official weather forecasts and charts available, not simply from newspapers and TV stations but particularly from the aviation reports and forecasts obtainable by phone, fax or computer. Due to reports and forecasts being subject to possible amendment in terms of time, layout and codes when used, they are not dealt with in detail in this book. How you can be sure you are up to date is covered later.

SOURCES OF AVIATION REPORTS & FORECASTS
THE UNITED KINGDOM

In the UK a full presentation of sources of weather reports, forecasts, and any codes involved, is available in an excellent free pocket-sized booklet from the Civil Aviation Authority (CAA) called *GET MET*. Copies of this booklet can obtained by sending an A5 size stamped addressed envelope to: Safety Promotion, General Aviation Department, Civil Aviation Authority, Aviation House, Gatwick Airport South, West Sussex RH6 0YR.

In fact, consider the *GET MET* booklet to be an essential supplement to this book for flying in the UK as questions on weather information sources and their content can be asked in your meteorological examination paper.

METFAX/METWEB

Apart from information available by phone, data from these two sources comes as their names imply – METFAX via a fax machine or METWEB via a computer through the Internet. These services embrace a considerable range of reports and forecasts including weather charts – the latter not being possible to obtain by phone. There are laid down charges and terms for their use.

AIRMET

Extensively used, this is a public telephone (or fax) service where regularly updated aviation forecasts are given out at dictation speed.

To maximise the use of AIRMET through the telephone, there is available a form known as the *Pilot's Proforma*. It includes a chart and is laid out in the order to which the forecast is given over the phone. Supplies of the proforma can be obtained from AOPA, 50a Cambridge Street, London SW1V 4QQ. You can also request an explanatory leaflet and user's notes. The forms and notes are free but you need to send an A4 size stamped addressed envelope marked *Pilot's Proforma* on the outside. Photocopying for additional copies of the proforma is permitted.

METAR and TAF

There are two important sources of information with which you will be expected to be familiar; they are the **METAR** and the **TAF**. They come to you in a code format agreed by the *International Civil Aviation Organisation (ICAO)*. The codes are internationally accepted apart from where a country has registered differences. The reference source, which will contain the very few differences in ICAO codes for the UK, is *Section Gen 3-5* of CAA publication *CAP 32 – The UK Aeronautical Information Publication (AIP)*. Again, when changes to codes and any other aspects of aviation reports/forecasts are made in the UK they are usually published in a *CAA – Air Information Circular (AIC)* as well as in the index pages of *GET MET*. So far as using METARs or TAFs is concerned, you are strongly advised to become familiar with at least the basic data found in same together with codes relevant to same – you could be tested on this point.

METAR (OR SPECI)

The *METAR* is an *actual* (current at the time stated) weather report from a given airfield carried out by a qualified meteorological observer and not necessarily by a forecaster. *SPECI* refers to a similar but non-routine report. Such a report is currently issued in the following sequence:

1	IDENTIFICATION	(a)	METAR or SPECI
		(b)	An ICAO four-letter station code allocated to the airfield. e.g. EGSS London Stansted
2	SURFACE WIND	(a)	Wind direction/speed in knots
		(b)	Extreme direction variance
3	VISIBILITY	(a)	Minimum visibility
		(b)	Maximum visibility
4	RVR		Runway visibility
5	PRESENT WEATHER		Groups of codes
6	CLOUD		Amount/type/base
7	CAVOK		Visibility >10km with No cumulonimbus cloud No cloud below 5000 feet *Note: When CAVOK is reported there are no visibility or cloud reports given.*

8	TEMPERATURE/DEW POINT	Both in degrees Celsius
9	QNH	Mean sea level pressure 'Q' indicates millibars – If 'A' is used it denotes inches
10	RECENT WEATHER	Groups of codes (max. of three)
11	WIND SHEAR	Currently not reported in the UK
12	TREND	Coded future weather
	An equal sign = indicates the end of a report	

TAF

The TAF (Terminal Aerodrome Forecast) is a forecast covering a specified airfield for a selected fixed period from the time issued. A TAF is regularly updated – subject to the forecaster having access to two consecutive (or at least one) METARs from that airfield on which a forecast can be based. Should an airfield be closed, the forecaster may well have to wait until it has been open for at least an hour to be able to obtain the required METAR.

A TAF report is issued in the following sequence:

1	REPORT TYPE	TAF
2	LOCATION	An ICAO four-letter station code allocated to the airfield. e.g. EGSS London Stansted
3	DATE/TIME OR ORIGIN	Zulu time (Z) which is Greenwich Mean Time (GMT). A 'Z' can sometimes follow the time figures.
4	VALIDITY TIME	Hours from – to
5	SURFACE WIND	Direction to nearest 10 degrees (T). (In circuit it will be in degrees (M) when provided from a control tower) W/V is invariably in knots e.g. 220/10, 030/25, etc.
6	VISIBILITY	Shown as a minimum in metres or kilometres or as CAVOK

7	SIGNIFICANT WEATHER	Anticipated weather in code
8	CLOUD	Types/Amounts in code. Height of base is in hundreds of feet above airfield level.
9	SIGNIFICANT CHANGES	(a) % probability at 30% to 40% (b) Time (c) Met Groups depicting changes in part or in total of weather.

When obtaining a TAF it is naturally logical to obtain a METAR for the airfield at the same time; they complement each other and need to be used jointly. Be careful when using METARs and/or TAFs if flying to a destination which is **not** the reporting airfield but happens to be one fairly close to same. Taking in account any hills, mountains or any coastal factor in the area; such terrain and proximity to the sea can dramatically produce different weather conditions between two locations only a few miles apart.

INFORMATION

Should it prove difficult or impossible to obtain any of the services provided, your copy of *GET MET* contains contact telephone numbers of use. Furthermore, should you need amplification of some forecasts or route forecasts not covered in the information available, the booklet gives you the numbers to contact. In using any of these numbers you must show evidence of having already obtained the basic information to hand. In the meantime here are some other services available.

VOLMET

Volmet is a VHF radio service operating on a number of frequencies providing METARs to aircraft in flight for a number of major airfields. It works through a sequence of 10 or so airfields every few minutes and is constantly updated.

FORECASTER CONSULTANCY

The days have gone in the UK when you could contact your nearest met station and receive individual attention and a personal forecast as and when you wanted it – at no cost. However, there are times when the available information or even any amplifications of same, as mentioned earlier, is insufficient for your needs. For these occasions you can make further personal contact with an aviation forecaster but the current cost is £17 exclusive of the cost of the phone call. Compared with the cost of running a car on a wasted trip to your airfield it will probably be far more cost-effective to spend a few pounds on obtaining a professional picture of the weather prior to setting off to fly.

NOTE

Again, to ensure the interpretation of reports and forecasts is based on up-to-date information always refer to:

- The Met Section of the current CAP 32 – UK Aeronautical Publication (UK AIP)

or

- Any relevant Aeronautical Information Circular (AIC)

or

- The index pages referred to in the latest edition of the *GET MET* booklet.

SOURCES OF AVIATION REPORTS & FORECASTS
THE UNITED STATES OF AMERICA

In the USA personal contact with weather centres remains very much in being – alongside automated weather information. In fact contact with weather centres is available in flight; this is particularly necessary to obtain msl atmospheric pressure readings along a route. Furthermore, which must be to the envy of UK pilots, all the services are free! This is a demonstration of the attention paid in the USA to flight safety where assisting pilots at all times, be it in the air or on the ground, with weather conditions is taken seriously.

METARs and TAFs are also available with their codes being gradually changed to the international terminology. However, there will still possibly differ at times from the UK ones. There is a need to ensure that no changes have been made to any reference source you may be using.

The range of services is as follows:

FAA Flight Service Stations (FSS or AFSS)

There are two types – the Flight Service Station (FSS and AFSS). The former is manually operated and the latter is automated. There is currently a change taking place where non-automated centres are in the process of becoming automated. The FAA Service Stations provide more aviation weather briefing service than any other US Government department. The pilot has the choice of listening in to the automated version of a forecast or contacting either the FSS or AFSS directly.

Transcribed Weather Broadcast (TWEB)

This is a continuous broadcast on selected low/medium frequency navigation facilities (190 to 535 kHz) and VORs (108.0 to 117.95 mHz).

Pilot's Automatic Telephone Weather Answering System (PATWAS)

A local-area forecast provided by some manual FSSs for an area 50 miles in radius from the station involved. PATWASs are prepared at selected intervals and available between 0500 and 2200 hours local time. PATWAS Centres are listed in the *Airport Facility Director (AFD)*.

Transcribed Information Briefing Service (TIBS)

Provided by an FFS, TIBS provides continual telephone information of both meteorological and aeronautical information during the same hours as the

PATWAS. Again, these centres are listed in the AFD.

En Route Flight Advisory Service (EFAS)

Also known as *Flight Watch*, this is a service provided by selected FSSs and AFSSs on a common frequency of 122.0 mHz and on discreet frequencies when flight is above 18,000 feet – the transition altitude. The *Flight Watch* (EFAS) specialist provides aviation weather information, critical to an *en route* pilot faced with hazards or unknown conditions. Being on a common frequency you may hear more than one at the same time. Naturally you would work on the loudest, probably nearest station.

AM Weather

AM Weather is a 15-minute weather programme on the air every day from Monday to Friday on more than 300 Public Broadcasting Television Stations. The data is prepared by professional meteorologists and is most useful to the pilot in deciding first thing in the day if flying conditions are 'Go' or 'No-Go'. All the services mentioned are free.

For the complete and very comprehensive range of services provided in North America, together with the current codes for METARs and TAFs, you should obtain an updated copy of the joint publication issued by the Federal Aviation Administration (FAA) and National Weather Service (NWS) called *Aviation Weather Services AC 00-45D*.

NOTE

Comments, queries and suggestions are welcomed by:

> The National Weather Service Co-ordinator
> AMA - 9, FAA
> Mike Mahoney Aeronautical Center
> PO Box 25082
> Oklahoma City
> OK 73125-0082

CHAPTER TWENTY-FOUR

RECAP

We have now covered our study of the atmosphere and the effect it can have on an aircraft in flight. It is worth recapping some of the main aspects to round off the study you have put into the subject.

DENSITY

- When your airfield QFE is *lower* than the ISA standard of 1013.25 mb (hPas) (29.92 in. Hg), the density will also be lower than the ISA standard of 1.225 kg per cubic metre – that is, below 100%.

- Likewise, when the temperature is *above* the ISA standard of 15°C (59°F), the density will be lower than 100%.

- As all aircraft limitations in their flight manuals are calibrated against the ISA standard you must expect a longer take-off and landing run – together with a faster ground speed both for taking off and landing when density is lower.

PRESSURE/WINDS

- With a pressure decrease comes a density decrease, so take care at a high-altitude airfield because the problems related to a lower density are the same.

- Is your field large enough to cope with possible low density problems caused by lower pressures or higher temperatures?

- Make sure you have the appropriate altimeter setting for each stage of your flight for safe separation from other aircraft.

- QFE is set on your altimeter when due to land or flying in the circuit.

- QNH is set just prior to take-off and *en route* is changed to the regional QNH *en route* or SPS (Standard Pressure Setting) as appropriate at the time.

- In the USA, except above transition level, the altimeter is set at the MSL pressure at the time. There is no change on entering or leaving the circuit — only when MSL pressure changes occur *en route*.

- In the UK, the Standard QNH (SPS) of 1013 mb (hPa) is set when passing through the relevant transition level to fly in accordance with the pressure altitude known as a Flight Level.

- In the USA there is only one transition altitude; it is at 18,000 feet.

- When the QNH (MSL) is lower than the SPS, the pressure altitude will be higher than the QNH altitude.

- When the QNH (MSL) is greater than the SPS, the pressure altitude will be lower than the QNH altitude.

- When approaching controlled airspace in the form of a CTA remember that you could infringe that CTA if your QNH setting is lower than the SPS of 1013 mb.

- A clockwise change in wind direction is called veering; an anti-clockwise wind change is backing.

- When you see a met chart with isobars close together remember that the winds can be strong.

- In the northern hemisphere winds blow clockwise round an anticyclone (high) and anti-clockwise round a depression (low). In the Southern Hemisphere it is the opposite.

- The winds at 2000 ft and above will have veered by some 25 to 30 degrees from the direction at the surface over land.

- With a strong wind blowing from your left side or a drift to your right, remember that, unless adjusted, your altimeter will tend to show a false increase in height as you continue *en route*.

- Note that air close to the surface will behave like water in a river with a ragged bed; it will be turbulent.

- Wind over mountains can provide lift on the windward side but create havoc on the leeward side.

- A mountain range can produce a wave effect which at one point can provide useful lift but at another point can lead to an unexpected descent. Lenticular clouds are an indication of such a possibility.

- Except on landing in virtually nil wind conditions ensure you have an appropriate reserve of airspeed to allow for gradient (vertical wind shear) effect.

- Near thunderstorms there can be a horizontal wind shear that can cause a sudden drop in airspeed on approach with the same effect on your aircraft as wind gradient.

- Under a thunderstorm a microburst can cause severe vertical and horizontal wind shear.

- Undue turbulence can produce severe strain on your aircraft structure so remember to reduce power to manoeuvring speed.

- Sea breeze effect can be the cause of considerable variations in wind direction/speed (vertical and horizontal wind shear) on approach and landing at a coastal airfield.

- Sea breezes can produce mini-cold frontal conditions inland which would not appear on a normal weather chart.

- If you launch from a hill-top in a hang glider at the end of the day and fail to become airborne – remember the katabatic wind.

TEMPERATURE

- The Tropopause is higher at the Equator than over the Poles. It is also higher in summer than in winter.

- The higher the temperature – the higher the Troposphere; the lower the temperature – the lower the Troposphere.

- The ISA lapse rate is fixed at 1.98°C with 2°C normally accepted.

- Temperature normally decreases with height at a variable rate known as the Environmental Lapse Rate (ELR).

- Air forced to rise will cool at a fixed rate:
  Dry Adiabatic Lapse Rate   (DALR)   at   3°C   (5.4°F)
  Saturated Adiabatic Lapse Rate (SALR)   at   1.5°C (2.7°F).

- When the ELR is less than the ALR – conditions are stable.

- When the ELR is greater than the ALR – conditions are unstable.

- A temperature increase with height is known as an inversion, it can trap haze, mist and fog.

- An unstable atmosphere can lead to varying degrees of turbulence during flight.

- Uneven surface heating in unstable conditions causes convection which, in turn, brings additional turbulence to that caused by surface winds. However, convection turbulence can also extend to any level in the Troposphere.

HUMIDITY

- Relative humidity will increase in cold air and decrease in warm air.

- The higher the relative humidity – the lower the density.

- Condensation from invisible water vapour into visible water particles takes place when the relative humidity exceeds 100% – given that condensation nuclei are present.

- When below dew point, at over 100% relative humidity, if no condensation nuclei are present to permit condensation to take place, the air becomes known as supersaturated air.

- When water vapour changes directly into ice crystals the process is known as sublimation.

CLOUDS

- Clouds form when moist air is raised to cooler levels by processes such as orographic, turbulence, convergence, frontal and convection.

- The base height of each class of cloud can vary with the location in the world and the seasons.

- Less turbulence found in stratiform than in cumuliform clouds.
- Precipitation and weather associated with clouds:

Drizzle:	Thick stratus and strato-cumulus.
Patchy rain/snow:	Thick strato-cumulus.
Continuous rain/snow:	Nimbo-stratus/dense strato-cumulus.
Cont's heavy rain/snow:	Thick nimbo-stratus.
Showers:	Large cumulus and cumulo-nimbus.
Severe turbulence:	Large cumulus and cumulo-nimbus.
Hail, thunder:	Cumulo-nimbus.
Squall, microburst:	Cumulo-nimbus.
Wind shear:	Cumulo-nimbus.
Structural damage:	Cumulo-nimbus.
Wave effect:	Lenticularis (all levels).
Poor visibility:	Haze, mist, fog and precipitation.

ICING

- Hoar frost accumulated on an aircraft must be cleared before any flight is attempted.
- Rime ice forms in clouds composed of relatively small water droplets.
- Clear ice forms in cu-nimb cloud where large super-cooled water droplets in abundance contact the aircraft and freeze in stages.
- Rain ice, also known as freezing rain, can form in clear air when rain droplets fall from alto-stratus at a warm front into freezing air below and become super-cooled. They freeze on contact with the cold surface of the aircraft.
- Carburettor icing can form in moist air at temperatures ranging from –7° to 21°C (20° to 70°F). Most likely during descent with the engine throttled back – hence the need for periodic bursts of power.

FRONTS

- Fronts are usually associated with polar depressions and are preceded by falling pressure on the barometer.
- The ana system can reach to high cloud level.
- The ana system sequence is cirrus, cirrus stratus, alto-stratus, nimbo-stratus with continuous rain/snow, strato-cumulus/stratus, nimbo-stratus with continuous rain or snow for shorter period and cirro-stratus with subsequent cu and maybe cu-nimb.

- During the cold front, nimbo-stratus in the above sequence can be replaced with cu-nimb with very heavy rain and even thunder if the air at the ana cold front is unstable.

- The kata front sequence is thickening strato-cumulus perhaps becoming nimbo-stratus in appearance if very thick with continuous rain/snow, easing to thinner strato-cumulus/stratus, thickening strato-cumulus with further but shorter period of rain/snow, thinning strato-cumulus with subsequent cu or even cu-nimb.

- Wind veers as each front passes through.

- Always fly towards a front if enjoying an hour or two in the air. This will leave the way home clear.

THUNDERSTORMS

- Usually occur at the end of a summer heatwave in the temperate latitudes, preceded in hot humid conditions by a build up of cu-nimb in the afternoon.

- May be preceded by alto-cumulus castellanus and/or floccus when forming at medium level.

- Can also occur in unstable cold fronts and occlusions.

- Keep well away!

VISIBILITY

- Fog in aviation practice is said to exist when visibility falls below 1 kilometre (3500 ft) just over half a mile.

- Given a clear sky, radiation fog is quick to form on an autumn or winter evening and can do so very early on a spring or summer morning. Beware of that attractive late afternoon clear sky during autumn and winter when you see evidence of high humidity in the form of condensation, say, on your car.

- Near coasts when the wind is off-shore blowing moist air into high ground, keep away from that high ground as low stratus can be prevalent.

- Remember, a vertical visibility enabling sight of an airfield from directly overhead may be insufficient to keep the airfield in sight on horizontal approach.

- Individual cloud shadows on the ground, given a sunny day, can easily make it difficult to identify an expected way-point.

- Remember, 'Global Positioning Second' – map, flight plan and 'eye-ball' first.

REPORTS/FORECASTS

- Make full use of the met services available.

- Become familiar with METAR and TAF codes.

- Ensure your sources of codes for METARs and TAFs are up to date.

- If at all in doubt after examining a routine report or forecast, consult one of the nominated met centres.

- Ensure you check that the information you have on weather reporting and forecast systems is up to date.

There are no doubt other points which have not been covered in this recap chapter; but to go any further would be duplicating the book.

If any of the above is new to you, it is a signal to return to the drawing board!

CHAPTER TWENTY-FIVE

GENERAL DATA

The following depicts how figures for the International Standard Atmosphere vary as height increases. The figures are only approximates: eg where near the surface, the pressure drop related to height is 30 ft per unit of 1mb (hPa) or 0.03 in.Hg, after say 20,000 ft it becomes around 50 ft per unit increasing yet again to 75 ft per unit at around 30,000 ft.

INTERNATIONAL STANDARD ATMOSPHERE

Height		Pressure		Temperature Degrees			Density	
feet	metres	mb	in.Hg	C	F	%	kg	lb
35,000	10,670	240	7.09	−55	−67	31%	0.368	0.237
30,000	9140	300	8.86	−45	−49	38%	0.490	0.290
25,000	7620	375	11.07	−35	−31	45%	0.551	0.344
20,000	6100	466	13.76	−25	−13	53%	0.649	0.404
15,000	4570	570	16.83	−15	5	63%	0.772	0.481
10,000	3050	700	20.60	− 5	23	74%	0.907	0.565
5,000	1525	840	24.81	5	41	86%	1.054	0.657
Surface		1013.25	29.92	15	59	100%	1.225	0.764
Mean Sea Level		mb (hPa)	in.Hg				kg Cubic Metre	lb Foot

Altimeter Settings	Standard Pressure Setting (SPS)	1013 mb/hPa
	or	29.92 in.Hg
	Mean Sea Level Pressure (QNH)	Variable
	Airfield Pressure (QFE in UK)	Variable

Lapse Rates	Standard Atmosphere	1.98°C per 1000 ft
	(to 36,000 ft)	2.00°C is acceptable
	or	3.56°F per 1000 ft
	Environmental (E.L.R.)	Variable
	Dry Adiabatic (DALR)	3.0°C per 1000 ft
	or	5.4°F per 1000 ft
	Sat. Adiabatic (SALR)	1.5°C per 1000 ft
	(Up to 6000 ft) or	2.7°F per 1000 ft

Density   Not readily measurable by an instrument as with pressure or temperature. Simply remember that **density decreases** as *pressure decreases* and/or *temperature* and/or *relative humidity increases.*

CHAPTER TWENTY-SIX

FINALE

We have come to the end of our journey through the sky. Now come the days of putting the basic knowledge into practice and gaining experience. As mentioned earlier, when you wish to advance to commercial or airline pilot status, or perhaps a Gold C with diamonds as a top sailplane pilot, there are numerous books to take you that stage farther.

Meteorology is a fascinating subject, not one to be dreaded. The sky is ever telling you a story; reading and acting upon it can bring great satisfaction – ask any sailplane, hang glider or paraglider pilot. Wherever you may be in the world the principles are the same – only the extremes of weather are likely to differ.

However, just like the sea, the sky can bite and does so at times. But the statistics are largely made up of inexperienced or bold(?) pilots where the former in the early days should have been a little more 'chicken' and the latter should have been more aware of the pitfalls.

Always act upon the principle of
WHEN IN DOUBT – DON'T!!

Make use of the available reports and forecasts – the cost of a phone call/fax or a chat is really worth it. Possessing a barometer (or better still a barograph if you can afford it) is a useful aid for relating any forecast bad weather to the actual conditions in your locality at the time.

**KNOW THE SKY
IN WHICH YOU FLY**

The real need to learn begins *after* you have attained your licence or certificate and the instructor is no longer beside or behind you, but a passenger can be. The weather is one facet of flying in which there is no end to the learning process.

Finally, do seek further training for a higher rating if you wish to fly in adverse conditions. Apart from probably being illegal to do otherwise, it simply makes for common sense.

INDEX

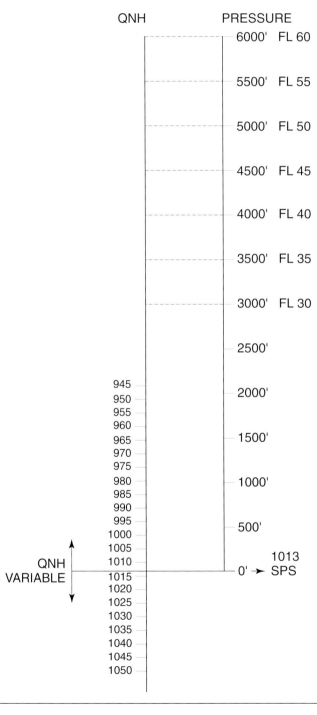

ALTITUDE SCALE

QNH   PRESSURE